PREHISTORIC NATIVE AMERICANS AND ECOLOGICAL CHANGE

Human Ecosystems in Eastern North America since the Pleistocene

Prehistoric Native Americans and Ecological Change shows that Holocene human ecosystems are complex adaptive systems in which humans interacted with their environment in a nested series of spatial and temporal scales. Using panarchy theory, it integrates paleoecological and archaeological research from the Eastern Woodlands of North America, providing a new paradigm to help resolve longstanding disagreements between ecologists and archaeologists about the importance of prehistoric Native Americans as agents for ecological change. The authors present the concept of a panarchy of complex adaptive cycles as applied to the development of increasingly complex human ecosystems through time. They explore examples of ecological interactions at the level of gene, population, community, landscape, and regional hierarchical scales, emphasizing the ecological pattern and process involving the development of human ecosystems. Finally, they offer a perspective on the implications of the legacy of Native Americans as agents of change for conservation and ecological restoration efforts today.

PAUL A. DELCOURT is a professor in the Department of Ecology and Evolutionary Biology at the University of Tennessee, Knoxville. His areas of research interest include reconstruction of ice-age environments and global climate change. Over his career, he has worked with archaeologists across southeastern North America to understand the relationships of prehistoric Native Americans to their changing environments.

HAZEL R. DELCOURT is also a professor in the Department of Ecology and Evolutionary Biology at the University of Tennessee, Knoxville. Her research has focused on the history of deciduous forest species from the Pleistocene to the present day. She is particularly interested in the application of insights from the paleoecological record as they can be applied to the conservation of biological diversity.

PREHISTORIC NATIVE AMERICANS AND ECOLOGICAL CHANGE

Human Ecosystems in Eastern North America since the Pleistocene

PAUL A. DELCOURT AND HAZEL R. DELCOURT

Department of Ecology and Evolutionary Biology
University of Tennessee, Knoxville

CAMBRIDGE UNIVERSITY PRESS
Cambridge, New York, Melbourne, Madrid, Cape Town, Singapore, São Paulo

Cambridge University Press
The Edinburgh Building, Cambridge CB2 8RU, UK

Published in the United States of America by Cambridge University Press, New York

www.cambridge.org
Information on this title: www.cambridge.org/9780521662703

First published 2004
This digitally printed version 2008

A catalogue record for this publication is available from the British Library

Library of Congress Cataloguing in Publication data

Delcourt, Paul A.
Prehistoric Native Americans and ecological change: a panarchical perspective
on the evolution of human ecosystems in eastern North America since the
Pleistocene / Paul A. Delcourt and Hazel R. Delcourt.
p. cm.
Includes bibliographical references and index.
ISBN 0 521 66270 2
1. Paleo-Indians – East (US) 2. Indigenous peoples – East (US) – Ecology.
3. Nature – Effect of human beings on – East (US) 4. Plant remains (Archaeology) –
East (US) 5. Paleoecology – East (US) – Holocene. 6. Biotic communities –
East (US) 7. East (US) – Antiquities. I. Delcourt, Hazel R. II. Title.
E78.E2D45 2004
304.2 – dc22 2003062730

ISBN 978-0-521-66270-3 hardback
ISBN 978-0-521-05076-0 paperback

To Jefferson Chapman, Dan and Phyllis Morse,
Jim and Cynthia Price, and Roger Saucier, valued mentors
and guides in our quest to understand prehistoric people
and their environments

Contents

Acknowledgements

We thank our friend and editor, Alan Crowden, for his insightful challenge to explore interconnections and interactions between prehistoric people and their environment.

Many colleagues offered enthusiastic encouragement, thoughtful advice, and key literature references and reprints to help cultivate this vibrant interface between archaeology, ecology, and paleoecology. We thank Marc Abrams, David Anderson, Kat Anderson, Bill Baden, Ed Buckner, Jefferson Chapman, Michael Collins, Wes Cowan, Gary Crites, James Dixon, Penny Drooker, Gayle Fritz, Donald Grayson, Kristen Gremillion, Stephen A. Hall, Cecil Ison, Richard Jantz, Jim Knox, Timothy Kohler, Mark Kot, Mark Lynott, Jim Mead, David Meltzer, George Milner, Dan and Phyllis Morse, Evan Peacock, Jim and Cynthia Price, Emily Russell, Roger Saucier, Gerald Schroedl, Theodore Schurr, Bruce Smith, Jan Simek, Lynne Sullivan, Ken Tankersley, Monica Turner, and Steve Williams.

PART I

Panarchy as an integrative paradigm

OVERVIEW

Whether or not prehistoric Native Americans were a significant ecological factor has been the subject of intense debate among generations of ecologists. During the latter half of the twentieth century, the pendulum of opinion has swung from the belief that American Indians had no influence on the composition and structure of plant and animal communities to the assertion that they were responsible for destruction of native habitats through over-exploitation of natural resources and widespread use of fire.

Quaternary paleoecology and archaeology are inherently multidisciplinary fields, drawing upon proxy information from data sets representing past states of climate, soils and geomorphology, biota, and material culture to make interpretations of changes in natural and cultural landscapes over millennial time scales. Time series of paleoecological and archaeological data yield chronologies of changes in both ecosystems and social systems. Before 1980, however, relatively few studies combined archaeological and paleoecological methods to determine objectively the kinds, extent, and duration of ecological influences by prehistoric Native Americans. Although a number of exemplary studies have been conducted since that time, the challenge remains to integrate available information effectively from paleoecology and archaeology in order to understand the linkages and dynamic feedbacks of processes underlying the documented patterns of change in prehistoric human ecosystems.

Panarchy theory has been developed recently by C. S. Holling and colleagues (Gunderson and Holling, 2001; Holling 2001) as an extension of hierarchy theory that includes cycles of adaptation in ecological and cultural processes. Panarchy theory is a heuristic explanatory model that views the development of human ecosystems as holistic, self-organizing, complex adaptive systems. Panarchy theory views the complex interactions between humans and their environment as adaptive responses that result in self-organized, hierarchical systems. In this book, we adopt panarchy theory as a contemporary paradigm for integrating paleoecological and archaeological data concerning the evolution of human ecosystems in North America during the Holocene interglacial interval.

In Chapter 1, *The need for a new synthesis,* we discuss the debate over the "myth of the natural man." We explain why a disparity of viewpoints has arisen concerning the role of prehistoric Native Americans as agents of ecological change, and we suggest why a new kind of synthesis is needed.

In Chapter 2, *Panarchy theory and Quaternary ecosystems,* we adopt a panarchical view of the development of Quaternary ecosystems as self-organizing, complex adaptive systems. We illustrate the utility of panarchy theory by applying it on a Quaternary time scale, where glacial–interglacial cycles represent macro-scale adaptive cycles of organization, disruption, and reorganization of ecosystems.

In Chapter 3, *Holocene human ecosystems,* we explain why the Holocene interglacial interval is unique, with the arrival of humans in North America and the development of a several-tiered panarchy of human ecosystems. We explore the implications of this new viewpoint of human ecosystems as self-organized, complex adaptive systems for reinterpreting the classic cultural periods identified in the archaeological record.

1

The need for a new synthesis

THE MYTH OF THE NATURAL MAN

The concept of the Native American as "noble savage," living in harmony with nature, stems largely from the writings of the French philosopher, Rousseau, who lived from AD 1712 to 1778. Rousseau contrasted the "natural man," who lived in the unspoiled wilderness, with the "civilized man," who lived in society and sought to conquer the wilderness (Dolph, 1993). The concept of the noble savage was used in describing Native Americans by European American colonists of the eighteenth and nineteenth centuries, and those early descriptions became the basis for subsequent interpretations of the "presettlement" North American landscape by environmental historians and historical ecologists (Cronon, 1983; Crosby, 1986; Russell, 1997; Krech, 1999). The prevailing viewpoint of early twentieth-century ecologists was that human interference with natural succession began with European American settlement, and that in prehistoric times the vegetation was in equilibrium with climate and other environmental factors, including aboriginal human presence (Clements, 1936).

The debate among ecologists

Day (1953) was among the first academic ecologists to question this assumption. He argued that at the time of first European contact, the landscape of Eastern North America was not an unbroken virgin forest that had not been disturbed significantly by activities of Native Americans. Rather, settlers often described large tracts of open forest with little understory. Day cited ethnographic accounts (e.g., Densmore, 1927) to infer that prehistoric and historic Indians of New England drew upon the forest for a large variety of products. Historical accounts documented that Indians lived in villages that were sometimes stockaded and more than 100 acres in extent, and grew maize, tobacco, beans, and squash in extensive fields that were abandoned only as white settlements encroached and disease eliminated Indians. Day cited a number of anecdotal accounts of the use of fire, for example by the Iroquois of central New York, for hunting, improving growth of grass, and clearing of fields. He suggested that Indians may have taken advantage of blowdowns of forest canopy trees, keeping the forest canopy gaps clear by burning. Day further speculated that selective hunting and use of fire to promote browse would have increased the populations of heath hen, passenger pigeon, wild turkey, and white-tailed deer, and that the cultural practice of favoring nut trees and other food plants would have had long-term effects on forest composition. He stated, however, that there was no evidence in the writings of early authorities for wholesale annual burning across southern New England, and that fire was important only in places inhabited by Indians. Day (1953) concluded that "Indians of the Northeast cleared land for villages and fields, cut fuelwood and set fires beyond the clearings, exercised a wide indirect influence on vegetation through their hunting, and may have favored or even transplanted food and medicinal plants."

Later researchers (Cronon, 1983; Russell, 1997) questioned the extent to which anecdotal evidence in the form of early travelers' accounts and diaries can be used to substantiate conclusions about the importance of Indians as an ecological factor. In a re-evaluation of the historical evidence for burning by Native Americans within the forested coastal region extending from the Carolinas to Maine, Russell (1983) found that only half of the thirty-five documents she read that described vegetation or Indian life in the sixteenth and seventeenth centuries mentioned the use of fire, except for cooking. Only six credible accounts referred to purposeful burning near camps or villages, with fires escaping only accidentally; the most frequently described fires were scattered and of limited extent. Russell cited an earlier study by Raup (1937) suggesting that early settlers may have attributed open woods to Indian-set fires because they were unable to imagine natural openings in the forest.

Although historical accounts can provide direct evidence linking human activities with ecological effects, such documents must be screened carefully to ensure objectivity of the witness (Forman and Russell, 1983). From analysis of such documents, the extent of the ecological effects of limited use of fire and other human activities remains unclear, with several alternative explanations for the described openness of precolonial forests: (1) descriptive accounts were written by land speculators to attract colonists, and emphasized how easily open woodland could be converted to farms; (2) settlements of European Americans were established on the sites of abandoned Indian villages, and travel routes followed long-established Indian trails, leading to a perception of widespread disturbance from a biased, localized view that may not have been representative of the vegetation across the whole landscape; and (3) shading by large, old canopy trees may have created an open understory in the forest through natural plant succession. Frequent Indian-set fires may have had deleterious effects on the environment, including destruction of necessary firewood and edible mast as well as replacement of predominant oak forests by fire-adapted pines (Russell, 1983).

Other authors have attempted to debunk what they called the "environmentalist myth" stemming from the romanticized concept of Rousseau's "noble savage" living in harmony with nature (Diamond, 1986; Dolph, 1993; Krech, 1999). Rather than revering nature and practicing a conservation ethic, prehistoric hunter-gatherers and primitive agriculturalists may have had widespread destructive influences on the environment, for example causing the extermination of large animals, whose extinctions occurred soon after first arrival of humans on the islands of New Zealand, Madagascar, and Hawaii. This "Overkill hypothesis" has been proposed to account for extinction of large mammals across North America at the end of the Pleistocene (Martin, 1984). Examples of irreversible environmental damage through habitat destruction associated with prehistoric human cultures include deforestation of Easter Island by Polynesian colonists that led to soil erosion, lower crop yields, loss of material for building canoes that were used for fishing, and loss of wood levers for erecting stone statues (Diamond, 1986). When the carrying capacity of the island for humans was exceeded, cannibalism and warfare resulted. In the American Southwest, a combination of prolonged drought, failure of irrigation systems, and over-exploitation of wood resources by the Anasazi led to a similar demise of their culture nearly 1000 years ago (Kohler, 1992; Redman, 1999).

Day (1953) called for increased documentation of the extent and intensity of effect of prehistoric Native American impacts on their natural environment. He advocated that to understand the ecology of a given study area, ecologists should collaborate with archaeologists to determine (1) the duration of Indian occupation; (2) prehistoric human population density; (3) population concentration and movements; and (4) the local pattern of settlement. He concluded "that an area which was wooded when first seen by white men was not necessarily primeval; that an area for which there is no record of cutting is not necessarily virgin; and that a knowledge of local archeology and history should be part of the ecologist's equipment." Despite Day's encouragement for interdisciplinary collaboration, until the

1980s and 1990s relatively few studies combined ecological, archaeological, and paleoecological methods to determine objectively the kinds, extent, and duration of prehistoric Native American ecological influences.

Divergent viewpoints

The continuing debate over the extent to which Native Americans have been an ecological factor through time has in part resulted from a lack of communication across disciplinary lines. This lack of coordination of interdisciplinary evidence has resulted from a long history of independent development of these academic disciplines. Across the North American continent, the fields of ecology, archaeology, and Quaternary paleoecology have developed parallel but distinctively different traditions of investigation based upon their respective long-established methods. During much of the twentieth century, ecological research concentrated on understanding the relationships of plants and animals to their natural environment, independent of human intervention (McIntosh, 1985). Archaeologists focused on interpreting the evolution of human cultures based primarily upon field collections of durable artifacts such as stone projectile points and ceramic pottery (Griffin, 1952). Quaternary paleoecologists have been preoccupied with deciphering long-term changes in climate from pollen records preserved in lake sediments (Wright *et al.*, 1993).

With these several divergent objectives, relatively little research has been designed to develop and test explicit hypotheses about the interrelationships of human populations, plant and animal communities, climate change, and landscape evolution. As a result, widely differing points of view have developed concerning the importance of prehistoric humans as agents of ecological change. Archaeologists, working with tangible evidence of human presence from numerous specific sites, argue for the pervasive impacts of Native Americans on North American landscapes throughout at least the past 15 000 years (Driver and Massey, 1957; Morse and Morse, 1983). Ecologists, selecting study sites located away from known centers of prehistoric human influence, view that in preColumbian times the North American continent was a vast, largely untouched wilderness (Braun, 1950; Williams, 1989; Whitney, 1994). Quaternary paleoecologists, using large lakes as study sites, strive to resolve regional climate change and broad-scale vegetation dynamics (Webb *et al.*, 1993), but tend to overlook ecological pattern and process on the more local scales at which humans may have interacted with their environment in prehistoric times (Delcourt *et al.*, 1986). Many paleoecologists thus view climate change as the ultimate forcing function for both ecological and cultural change (Wright, 1984, 1993), whereas many archaeologists argue for human cultural and technological adaptation as an overriding factor in social organization and population growth through time (Johnson and Earle, 2000; Kohler *et al.*, 2000).

EFFECTIVE LAND MANAGEMENT REQUIRES A NEW SYNTHESIS

Differences in techniques, focus, and objectives have resulted in a polarization of viewpoints, leading to confusion about the ecological role of prehistoric Native Americans. Resolving this conflict of opinions is important not only because it influences how we view the heritage of Native Americans but also because it is an important element in determining how we will manage our natural resources into the future (Peacock, 1998). The United States federal government mandates that wilderness areas and old-growth forests be restored and maintained in their "natural condition," defined as that which existed in preColumbian times (before European American contact in the late 1400s), and therefore supposedly in a condition that was previously unaffected by human activities (Henderson and Hedrick, 1991; Hamel and Buckner, 1998). This land management policy reflects the widely held notion that activities of prehistoric Native Americans were an insignificant ecological factor, and it is furthered by the general rejection by ecologists of "soft" data from archaeological investigations and historic ethnographic accounts (Russell, 1983, 1997).

Intermeshing of viewpoints

Integrating interdisciplinary data first requires intermeshing of viewpoints. Butzer (1982) was among the first to call for such a synthesis of viewpoints, with a general goal of understanding the human ecosystem. Stoltman and Baerreis (1983, p. 252) defined the human ecosystem as "a discrete human population that has a shared cultural inventory of technologic, social–organizational, and subsistence practices and is in interactive association with a specified environment." The human ecosystem thus defined includes subsistence, technology, social organization, population, and environment (Figure 1.1), all five of which potentially leave interpretable physical traces in the archaeological record (Stoltman and Baerreis, 1983). Butzer (1982) defined "contextual archaeology" as "a realistic appreciation of the environmental matrix and of its potential spatial, economic, and social interactions with the subsistence–settlement system." Thus, to an archaeologist, the environment is a "potential resource catchment" from which people draw

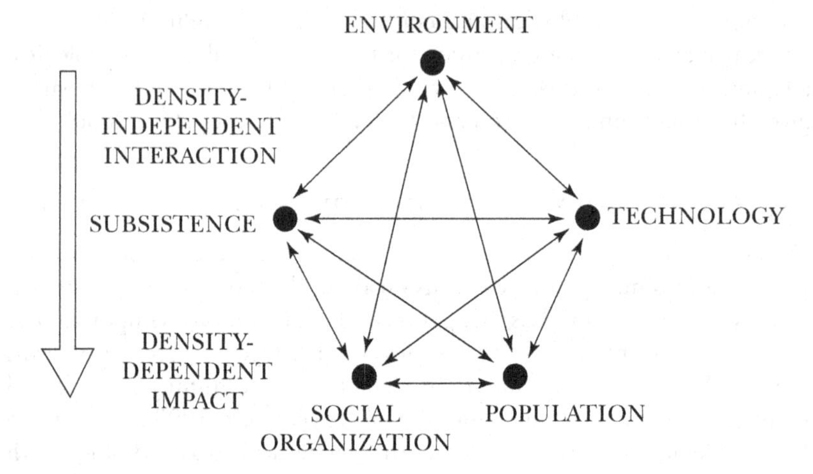

Figure 1.1 Model of the human ecosystem displaying the spectrum of human–environment interactions that reflect increasing impacts associated with larger populations (modified from Stoltman and Baerreis, 1983).

their sustenance. This viewpoint differs substantially from that of the paleoecologist, who views the activities of humans as one of many potential influences on ecosystem patterns and processes (Delcourt and Delcourt, 1991). To achieve integration of the reciprocal relationships between humans and environment, researchers from each discipline must be open to new viewpoints concerning the implications of their data. Only by integrating ecological, archaeological, and Quaternary paleoecological information can alternative hypotheses concerning the interrelationships of prehistoric Native Americans with their environment be tested.

Perceived obstacles to interpretation

A major reason for continued debate on the ecological effects of prehistoric Native Americans is the increasing rigor required for hypothesis testing. Perceived obstacles to achieving an effective synthesis of paleoecological and archaeological data include several traditional stumbling blocks in scientific investigation (Birks, 1988).

The first of these obstacles is that time series of fossil pollen or of cultural artifacts document unique sequences of events at particular places and may not be replicable as "natural experiments" (Deevey, 1969).

Second, paleoecological studies are primarily descriptive accounts of changes in landscapes through time, and thus interpretations may be difficult to present as falsifiable hypotheses; instead, multiple working hypotheses are necessary because several interpretations may be consistent with available evidence (Birks, 1988). In addition, a reductionist view of nature, in which a single factor is isolated as a cause of change, is untenable because many environmental and biological variables change simultaneously at a given location (Gunderson *et al.*, 1995; Abel 1998). Hence, as the demand for quantitative and reproducible data has grown, so has the realization that statistical correlations and determination of cause and effect are made more difficult by real-world circumstances where the paleoecological investigator cannot always separate environmental causes of ecological change from the effects of human cultural evolution.

Interdisciplinary integration also requires becoming open to new sources of data. Intermeshing of data sets reaches its highest fruition as a result of collaboration on data analysis (Griffin, 1984). Because of the tendency for specialization along disciplinary lines (Butzer, 1975), developing a holistic integration of data requires interpretation of many lines of evidence simultaneously, following discussion from several different sources of informed opinion.

Scaling issues

Correct interpretation of the interrelationships between humans and their environments requires an appropriate choice of spatial–temporal window through which prehistory is viewed (Delcourt *et al.*, 1983; Delcourt and Delcourt, 1988). To be effective, an interdisciplinary research strategy must take into account how humans view their environment, and at what scales in space and time their activities have influenced ecosystem development. We can begin to test hypotheses about human–environment interactions only if we can match the spatial and temporal scale of archaeological evidence for past human activities with that of paleoecological evidence for ecological changes. Determining the appropriate spatial and temporal scales (Levin, 1992) and choosing the appropriate analytical tools with which to compare archaeological and paleoecological evidence (Delcourt *et al.*, 1983) is the key to integrating these traditionally disparate forms of data.

Three discrete scaling issues are important in detecting human influences on ecosystem pattern and process: (1) determining the "grain" of landscape patterning in ecological and human cultural systems; (2) establishing the appropriate extent of the geographic area to be sampled by the different techniques; and (3) finding the appropriate temporal and spatial scales at which to detect interrelationships between humans and their environments.

By choosing appropriate techniques and sampling sites, optimal scales can be chosen for integrated paleoecological and archaeological studies. For example, alpha diversity (species richness, Whittaker, 1972) and local human impacts can be studied by combining pollen, plant macrofossil, and charcoal particle studies from soils and barrow pits with artifacts, ethnobotanical remains, and faunal remains from individual features and structures (microscale analysis). Beta diversity, or community composition across environmental gradients on watersheds (Gauch, 1982) can be studied through paleoecological analysis of sediments from ponds or small lakes located within tens of meters to a kilometer from an archaeological site representing a village or ceremonial complex (mesoscale analysis). Gamma diversity, or the patchwork of the ecological mosaic across a broad, regional landscape (Whittaker, 1972), can be determined from paleoecological studies of medium-sized to large lakes combined with analysis of integrated site systems across a large region (macroscale analysis).

CONCLUSIONS: FORGING A NEW SYNTHESIS

Several trends in research directions in the late twentieth century and early twenty-first century are setting the stage for developing a new synthesis forged from the extensive literature in ecology, archaeology, and Quaternary paleoecology. Landscape, historical, and restoration ecologists are increasingly focused upon the issue of future biological sustainability (Lubchenco *et al.*, 1991). Increasingly, ecologists are gaining an appreciation of the role of traditional agriculture in the cultural history and sustainability of human ecosystems both in Europe (Birks *et al.*, 1988) and in North America (Turner *et al.*, 2000). In addition, information gleaned from oral histories and ethnographic accounts is leading to a better understanding of human–environment interactions based upon traditional ecological knowledge (Berkes *et al.*, 2000). The legacy of prehistoric Native American ecosystems for future land management can be understood through synthesis of the development of such ecological interactions over time scales of hundreds to thousands of years.

2

Panarchy theory and Quaternary ecosystems

PANARCHY THEORY

The term "panarchy" was coined as a combination of "pan," the ancient Greek god of nature, and "hierarchy" (Holling, 2001). Panarchy theory is an extension of ecosystem theory, which rests upon the assumption that a hierarchically nested series of adaptive cycles result in self-organizing, evolving, complex natural systems that assemble across all scales in space and time. Holling (1995) defined four ecosystem functions that are interconnected by a series of flows of nutrients and energy (Figure 2.1). Ecosystem development begins with an exploitation phase during which opportunistic, pioneer species (r-strategists) dominate. During successional development, the ecosystem enters a conservation stage in which nutrients are garnered and biomass is gained, with early-successional pioneer species gradually replaced by late-successional species (K-strategists). The exploitation stage proceeds slowly to the conservation stage. Once attaining the conservation stage, ecosystems are vulnerable to more rapid changes that may be triggered by stochastic events such as fire, windstorm, or pathogen outbreaks, resulting in release of energy and nutrients (omega phase) and necessitating a major reorganization (alpha phase) during which the system can either recycle along the same loop or flip to an alternate state. This view of ecosystem development is consistent with results from long-term ecological research studies such as Hubbard Brook Watershed (Bormann and Likens, 1979). It forms an adaptive cycle that continually moves through the four phases of ecosystem growth, accumulation, restructuring, and renewal, but also is subject to evolutionary change (Holling, 2001).

Panarchy theory is predicated on the existence of a hierarchically nested series of such adaptive cycles, each with its own range of spatial–temporal scales. The hierarchy of ecosystems ranges across a wide spectrum of biological organization as well as suites of important processes (Allen and Starr, 1982; Delcourt *et al.*, 1983; Urban *et al.*, 1987; O'Neill *et al.*, 1986).

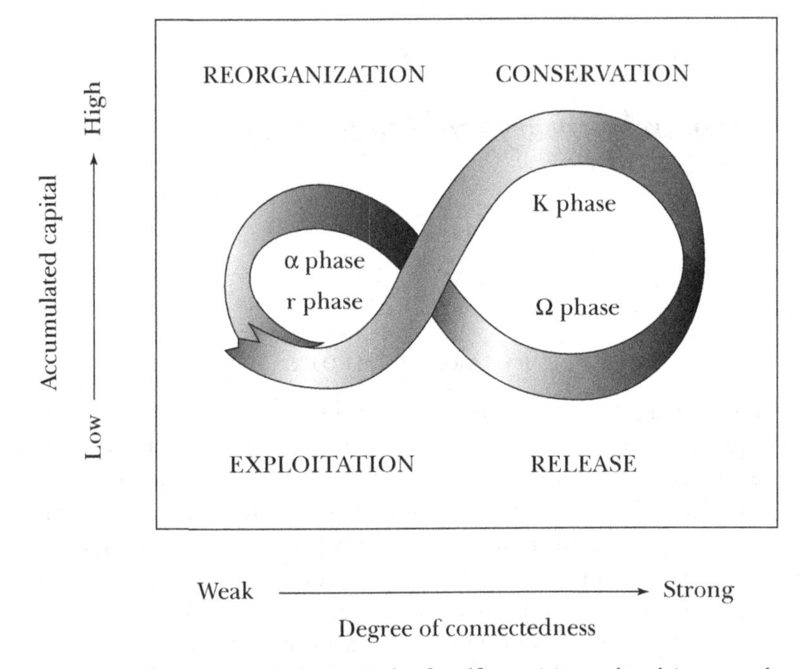

Figure 2.1 The panarchical adaptive cycle of a self-organizing and evolving, complex ecological system, involving four sequential phases of exploitation, conservation, release, and reorganization. This "Holling loop" is portrayed on an X-axis reflecting degree of connectedness and on a Y-axis for the amount of accumulated capital (for example, nutrients or biomass) stored within key ecosystem components (modified from Holling, 1995).

Each level in the hierarchy is a self-organizing system in which different initial conditions are channeled into similar final states because of positive feedbacks between structure and processes that mutually reinforce each other (Perry, 1995). The functioning and linkages between adaptive cycles in the nested hierarchy determine the overall sustainability of a panarchical system. The system may become increasingly vulnerable to nonlinear threshold effects within certain stages, such as the reorganization phase, in the adaptive cycle. Panarchical systems may collapse if subjected to major stochastic events such as rapid climate change or over-exploitation of resources (Holling, 2001).

THE QUATERNARY PANARCHY

Panarchy theory enables an understanding of both short-term (hundreds of years) and long-term (thousands to tens of thousands of years) ecosystem dynamics. Through the Quaternary Period of geologic history (the past two million years), terrestrial ecosystems evolved in response to broad-scale, periodic fluctuations in climate (Imbrie and Imbrie, 1979). With each of some 20 successive glacial–interglacial cycles, terrestrial ecosystems progressed through long-term cycles of consolidation during relatively stable glacial or interglacial intervals, alternating with destabilization and biotic reorganization at the transitions from glacial to interglacial conditions. As critical thresholds in climate tolerances of species were exceeded, over time global climate change has resulted in ecosystem development as a series of alternate stable states that can be interpreted to represent self-organizing adaptive systems. Ecosystem trajectories along these adaptive cycles constitute a panarchy of nested adaptive cycles of complex systems, separated by critical thresholds that define fundamental boundary constraints. These boundary constraints were described as "space–time domains" by Delcourt *et al.* (1983).

Cycles of global climate change

Through the past two million years, changes in the elliptical orbit of the Earth around the Sun, as well as in the tilt and wobble of the Earth on its axis, have generated periodic changes in the annual amount and seasonal contrast of incoming solar radiation. These inputs change predictably and drive the climate system on several independent "Milankovitch" cycles of 100 000, 41 000, and 21 000 years (Bennett, 1997). When superimposed on one another, these three cycles amplify or dampen their combined climate signal, resulting in rapid shifts between glacial and interglacial modes of global atmospheric circulation (Stuiver *et al.*, 1995). During the last 900 000 years, the solar pacemaker has set the planetary stage for the dominance of the 100 000-year cycle, with glacial conditions of 90 000 years' duration followed by interglacial conditions lasting on average 10 000 years. The transition from glacial to interglacial mode is relatively abrupt, with rapid global warming occurring over only a few thousand years; the transition from interglacial to glacial mode is more gradual, with build-up of continental ice sheets requiring tens of thousands of years. For most of each glacial interval, massive mountains of glacial ice mantle the continental landscapes of middle and high latitudes of North America, Greenland, and Europe, as maritime shelf ice and pack ice extend across the North Atlantic and North Pacific Oceans. Only during minor interstadial glacial retreats and major interglacial ice-free conditions are extensive deglaciated landscapes available for colonization by plant and animal species.

Glacial–interglacial adaptive cycles

Each glacial–interglacial climate cycle drives a macro-scale adaptive ecosystem cycle (Delcourt *et al.*, 1983; Holling, 2001). Watts (1988) defined an interglacial as an episode within the Quaternary of about 10 000 years' duration, during which soils, climate, and biota resemble conditions of today at the same latitude and longitude. Each interglacial period is initiated by a summer maximum of solar radiation and melting of ice sheets, followed by a protracted development of vegetation in response to continued climate change. Interglacials thus begin in the portion of the 100 000-year Milankovitch cycle when solar insolation is greatest and seasonal contrast is at maximum; they end when seasonal contrast is lowest and mean annual temperature is declining and ice sheets are rebuilding.

Based upon paleoecological records, the classic conceptual model of glacial–interglacial ecosystem development for the northern temperate zone proposes predictable, cyclic patterns of species invasions and terrestrial vegetation change (Watts, 1988). The concept of cyclic development of soil and vegetation in response to the glacial–interglacial climate cycle was first developed for Denmark and northern Europe by Iversen (1958), and subsequently extended to North America by Kapp (1977). Iversen (1958) defined four ecosystem phases, as follows (Figure 2.2): (1) a cryocratic phase during the time of rapid melting of continental glaciers, with newly deglaciated, frost-disturbed mineral soils colonized by steppe or tundra vegetation in full sunlight; (2) a protocratic phase during the time of increasing temperature in the late-glacial and early-interglacial intervals, with stable but unleached soils and with pioneer plant communities becoming invaded by woodland species in increasing shade; (3) a mesocratic phase during the mid-interglacial time of peak warmth, with brown, slightly leached soils developing beneath closed-canopy deciduous forest; and (4) an oligocratic phase during a time of deteriorating (cooling) climate, with acidic, podzolized soils and more open forest with heath and moorland. Three concepts are embedded within this conceptual scheme: (1) each interglacial represents a cycle in mean annual temperature, with an increase to an optimum, then a decrease; (2) cyclic changes in vegetation reflect the temperature cycle, progressing from open pioneer communities to closed late-successional forest, to open forest and heath; and (3) during the course of an interglacial, leaching of the soil is unidirectional, with soil pH changing from basic to neutral to acidic – with glaciation the parent material is renewed, setting the stage for the onset of the next interglacial cycle.

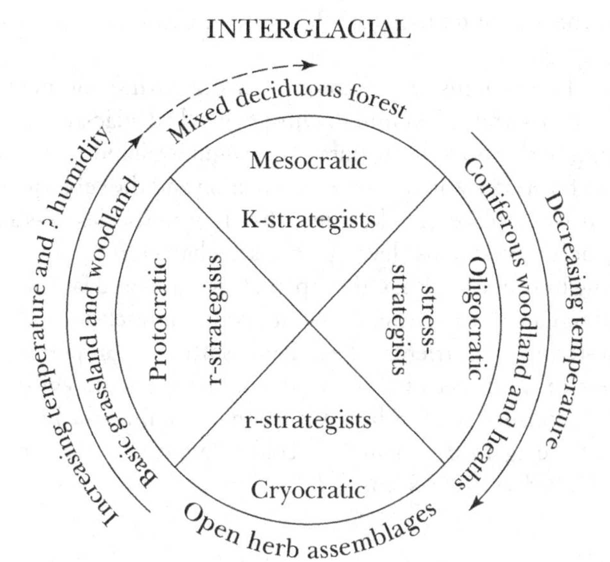

Figure 2.2 Iversen–Kapp model of the glacial–interglacial cycle of terrestrial ecosystem development for middle and northern latitudes of the Northern Hemisphere (modified from Iversen, 1958).

Kapp (1977) suggested that pollen records from midwestern North America also show predictable patterns of vegetation change. He proposed a five-stage cycle of glacial–interglacial vegetation phases for the Great Lakes region: (1) tundra or open boreal forest dominates during late-glacial times; (2) mixed conifer–northern hardwoods invade during early-interglacial times; (3) mesic deciduous forest, or prairie, dominate during peak interglacial warmth; (4) depauperate oak–hickory and spruce–heath communities replace mesic vegetation during the time of climate cooling; (5) ice, bogs, and tundra vegetation characterize full-glacial landscapes near the glacial–ice margin.

During glacial times, vegetation is relatively stable but organized in communities unlike those of interglacial times (Birks, 1986; Williams *et al.*, 2001). With the rapid transition from glacial to interglacial climate, these communities are in disequilibrium and are disassembled. A period of biological reorganization follows, during which stochastic events such as proximity of a colonization site to the former refuge area of a species or changes in seasonal contrast of temperature determine the initial order of community assembly.

Interglacial ecosystems are inherently self-organizing complex systems. Pioneer species (r-strategists) initially invade newly deglaciated landscapes, encountering fresh mineral substrates, high light availability, and minimal competition for space. With postglacial warming and the subsequent arrival and competitive displacement by longer-lived forest species (K-strategists), mid-interglacial ecosystems develop that are characterized by conserving and recycling nutrients within self-perpetuating, mid- and late-successional communities. Late in each interglacial interval, soil leaching and the long-term sequestering of nutrients in peatlands shift the competitive balance toward stress-tolerant species of trees and bog shrubs (s-strategists) occupying sites of diminished soil fertility. With climate cooling, the biotic systems reorganize, once again dominated by cold-tolerant pioneer communities (Kapp, 1977; Birks, 1986; Watts, 1988).

Interglacial ecosystems as repeated natural experiments

Pollen sequences from organic deposits preserved between glacially deposited till layers show that successive interglacial intervals represent repeated "natural experiments" (Wright, 1977). Similarities in the cycle of vegetation development exhibited from one interglacial to the next are evidence of the self-organizing nature of these complex adaptive systems. Glacial melting and retreat set the ecological opportunity for invasion. The magnitude and rate of interglacial climate change modulate the prevalent disturbance regime and influence the roles of invading species in structuring the assembly of communities. During each interglacial interval, cohorts of r-strategists, K-strategists, then s-strategists (Grime, 1979) compete for access to space, soil, nutrients, and light. The general sequence of invasion of species is conditioned by their individual adaptations to variations in seasonality of climate (Watts, 1988). The specific arrival sequence of species, however, is also determined by stochastic events including the chance dispersal of propagules, as well as the tendency for the resident community to exhibit biological inertia to invasion (Watts, 1973; Von Holle *et al.*, 2003).

Individual differences occur in the sequence of immigration of species on to deglaciated landscapes with each "shuffling of the ecological cards" (Wright, 1977). The order of species assembly, in addition to the nature of fine-scale disturbance events, influences which quasi-equilibrium state is attained (Grimm, 1983; O'Neill *et al.*, 1986; Birks, 1986).

CONCLUSIONS: THE PALEO-PANARCHICAL PERSPECTIVE

During each glacial interval, ecosystems undergo consolidation but are structurally, functionally, and compositionally different from those of interglacial intervals (Williams *et al.*, 2001). During late-glacial times of environmental instability and rapid climate change, these ecosystems are vulnerable to reorganization and exhibit nonlinear responses, or threshold effects (Birks, 1986). The resulting interglacial cycles of vegetation development can thus be interpreted as a suite of panarchical adaptive cycles of organization and reorganization, embedded within a longer-term Milankovitch glacial–interglacial climate cycle. The relative resilience or vulnerability of such complex adaptive systems is shaped by their geographic location with respect to latitude, the temporal context within the trajectory of climate change, and the presence or absence of anthropogenic influences.

3

Holocene human ecosystems

Panarchy theory provides a means of integrating ecosystem theory with sociology to explain the development of human ecosystems as complex adaptive systems. Panarchy theory views the complex interactions between humans and their environment as adaptive responses that result in self-organized, hierarchical systems. Human ecosystems are distinguished by the ability of people to use foresight, perform intentional actions, communicate ideas and experience, and make technological innovations. Currently, panarchy theory is applied primarily to issues of biological sustainability relevant to the present millennium (Gunderson and Holling, 2001).

Quaternary scientists and archaeologists examine environmental, ecological, and cultural changes on a variety of spatial and temporal scales. Panarchy theory is a powerful heuristic model that can provide a contemporary explanatory framework for integrating paleoecological and archaeological data over a nested series of hierarchical levels. This perspective provides a new paradigm for interpreting multiple forms of evidence concerning the processes that have underlain the development of late-Quaternary human ecosystems.

Humans have interacted with their environment since their arrival on the North American continent some 15 000 calendar years before present (BP, where 0 BP is equivalent to AD 1955, based on CALIB 4.3 calibration of radiocarbon ages; Stuiver *et al.*, 1998). The degree to which the actions of human beings resulted in ecological change has been determined by many factors, including human population size, lifeways, and the susceptibility of natural landscapes to anthropogenic alteration (Delcourt, 2002). We propose the following set of premises upon which to build a contemporary bridge between paleoecology and archaeology:

1. In the late Pleistocene and Holocene, *Homo sapiens* played an important ecological role as a keystone species, and prehistoric Native Americans had their own organism-centered view of landscapes and capability of ecological interaction and manipulation (O'Neill, 2001).

2. Prehistoric human ecosystems were nonunique outcomes of self-organizing adaptive development (Kohler *et al.*, 2000; Holling, 2001).
3. Prehistoric human ecosystems were metastable, nonequilibrium adaptive systems that changed continuously (O'Neill, 2001).
4. Through time, adaptive shifts in human ecosystems occurred as driven by nonlinear feedback loops, and their location on the cycle of ecosystem organization determined how vulnerable the human ecosystem was to change or collapse at a given time (Redman, 1999; Johnson and Earle, 2000; Holling, 2001).
5. Since the late Pleistocene, levels in the panarchy of North American human ecosystems have been added, reflecting temporally successive "cultural experiments" for which adaptive transformations coincided with the passage of critical thresholds in climate and environment, biota, or culture (Holling, 2001).
6. Late-glacial and postglacial aggregation of additional panarchical levels represented increasingly complex social organization in which new transforming strategies in ecosystem management were implemented for subsistence, shelter, and production of valued goods (Westley, 1995).
7. Trends in human population growth, dispersal, and occupancy of available territories have been linked to increasing social stratification in individual prestige and influence, specialization in work effort, religious–political governance, and enhanced connectivity with information transfer and spread of technological innovation throughout expanding trade networks (Anderson, 2001).
8. A panarchical view as a holistic, heuristic model is critical for placing observed past human ecosystem states into a meaningful framework of spatial–temporal processes and linkages between ecological and cultural change during the Holocene interglacial interval.

ANTHROPOGENIC ADAPTIVE CYCLES

The transformation of natural ecological systems to culturally managed ecosystems began in the late Pleistocene and continued through the Holocene. This transformation represents the cumulative development of adaptive management strategies, beginning with founding populations of the earliest Americans, who arrived in North America by at least 15 000 BP (Meltzer, 1993; Dixon, 1999; Bonnischen and Turnmire 1999a; Dillehay, 2000; Anderson, 2001; Adovasio and Page, 2002). The geographic pattern of iterative human impact is tied to their multiple entry times and routes for invasion, their colonizing dispersal and pattern of settlement, and the landscape-level repercussions of their population growth and innovations in technologic capability. With the "intertwining of ecological and social systems" (Westley, 1995), these interactive New World human ecosystems constituted a dynamic and complex adaptive system, coevolving in response to both the nature of human intervention and the ever-expanding scale of their influence (Holling, 1995).

Adaptive social systems

Westley (1995) incorporated the Holling heuristic model of adaptive cycles as the core insight for understanding social cycles of change, arguing that the emergence of a new adaptive social system in the r-strategist phase of revolution and exploitation (Figure 3.1) often occurs through the leadership of a charismatic visionary, who provides the catalyst for the cultural acceptance of a new vision or myth for the new social structure. These encoded beliefs may shape acceptance of adaptive strategies for managing ecosystems for greater security, improved shelter, and more predictable and/or augmented food reserves. Adaptive management translates human perception of resource availability and safe havens, as well as the physical pathways

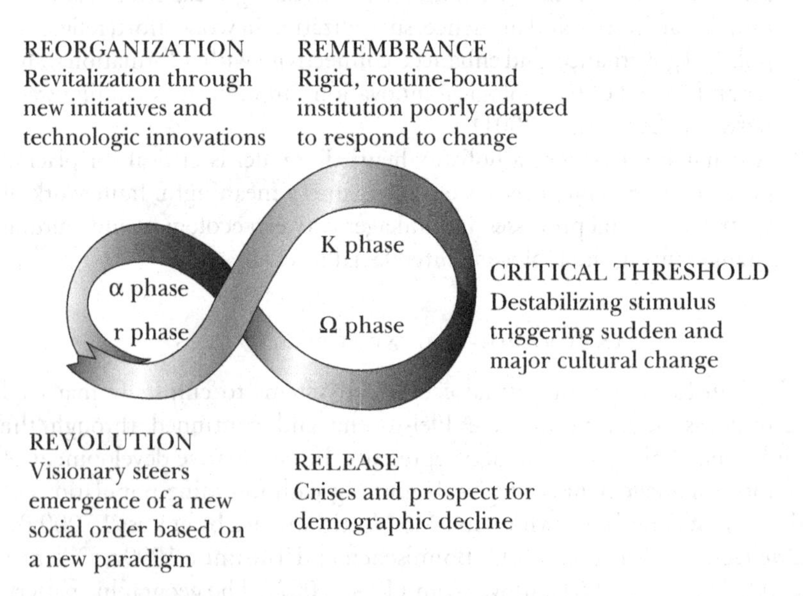

Figure 3.1 Revitalization of Adaptive Social Systems, with successive stages of revolution and exploitation (r phase), then remembrance and consolidation (K phase), release and restructuring (omega phase), and reorganization and renewal (alpha phase) (adapted from Wallace, 1966, and Wesley, 1995).

connecting resource patches, into a knowledge base used for weighing risk and reward. Cultural perceptions of the environment (world views, myths, paradigms) frame the typical spectrum of population activities bound by the collective rules and norms that govern routines. These "mental maps" model reality based upon past experience, traditional ecological knowledge, and then-current assumptions. Thus empowered, individuals are capable of recognizing changes within their environment, planning alternatives, making decisions, and allocating personal energies and material in order to implement responsive action. Thus, "learning provides an alternative to crises." For human groups "embedded" within their ecosystem, refining adaptive "strategy [serves] as a more natural, emergent process, the end result of the [cultural] creation of meaning within the [human eco]system and of ongoing learning linked to these meanings" (Westley, 1995, p. 394).

Remembrance eventually replaces revolt. The initial cyclic r phase of exploitation evolves into the organizational consolidation of communities and societies for greater efficiency (K phase), implementing more conservative strategic plans to ensure predictability and greater authoritative control over vital resources. With time, the increasingly rigid, routine-bound institution closes off to novel input and becomes less responsive to change.

During the omega phase of release and restructuring (Figure 3.1), destabilizing stimuli can exceed the cultural threshold for adaptability, triggering crises as impetus for sudden and cataclysmic change. Collapse or revitalization through the learning process may accompany technologic innovations or behavioral initiatives, creating new perceptions of opportunity.

The alpha phase of reorganization and renewal (Figure 3.1) incorporates the cultural legacy of traditional ecological knowledge as the foundation for implementing innovative solutions to societal dilemmas. Thus, adaptive social systems cycle through successive stages of revolution, remembrance, release, and revitalization (Wallace, 1966; Westley, 1995).

Individuals learn from personal experience, observations of others, and the accumulated base of traditional knowledge that creates meaning for interpreting ongoing experiences. Social units reinforce behavioral patterns through reward and punishment. Individuals and their communities learn from the environment, and can plan to act responsively to dynamic changes within the environment. Social organizations establish the authority structure for planning and decision making, as well as garnering and allocating resources. Social units of organization that become routine-bound and less responsive to challenges may be displaced, replaced, or eliminated. Social groups may reorganize through initiatives altering behavior to resolve problems or through innovations that, once created, generate opportunities (Parson and Clark, 1995).

Evolution of human societies

Johnson and Earle (2000) proposed an ecologically based, adaptive model for the evolution of human societies, tracking iterative feedbacks among the growth of human populations, technological development, and dynamically changing environments. Johnson and Earle (2000) maintain that the subsistence needs of expanding human populations invariably lead to collective actions for more extensively modifying and managing natural ecosystems, often with ecological costs in terms of habitat degradation. With their populations confronted by variable and limiting critical resources, individuals explore creative options to resolve shortfalls in food and shelter, as constrained by their existing technology and available natural resources.

This model for understanding the causes, mechanisms, and patterns of social evolution is based upon three kinds of cultural evolutionary processes: (1) subsistence intensification; (2) political integration; and (3) social

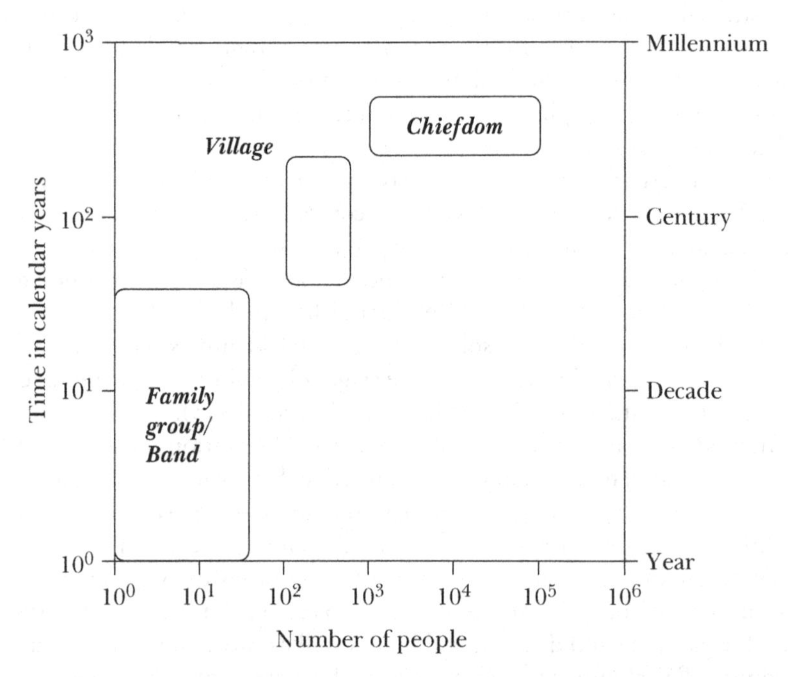

Figure 3.2 Three levels of social and political organization within which prehistoric people met their subsistence needs: (1) hearth-based family groups living in seasonal base camps; (2) local groups occupying villages for communal food storage and mutual defense; and (3) regionally integrated chiefdoms (adapted from Johnson and Earle, 2000).

stratification. Johnson and Earle (2000) proposed a synthetic classification of broad organizational units employing distinctive adaptive solutions for ensuring the health and safety of their members. How people met their basic subsistence needs can be understood at three levels of social and political organization: (1) hearth-based groups of several nuclear families living in seasonal base camps; (2) autonomous local groups occupying villages for communal food storage or mutual defense; and (3) regionally integrated polities such as chiefdoms and nation states (Figure 3.2).

The basic subsistence group of hunter–forager societies is the household unit of the nuclear family, comprised of five to eight people. Multiple families gather together seasonally, forming camps of kinship-based clans numbering 25 to 50 individuals (Figure 3.2). These ephemeral bands forage opportunistically for wild plant foods and hunt game on an annual round (Figure 3.3), with logistically organized forays across watersheds as large as 7×10^3 km^2 (Morse and Morse, 1983). The ecological impacts of hunter–foragers can be viewed as concentric circles radiating outward from their camp sites. Within a radius of up to three kilometers, a central core area

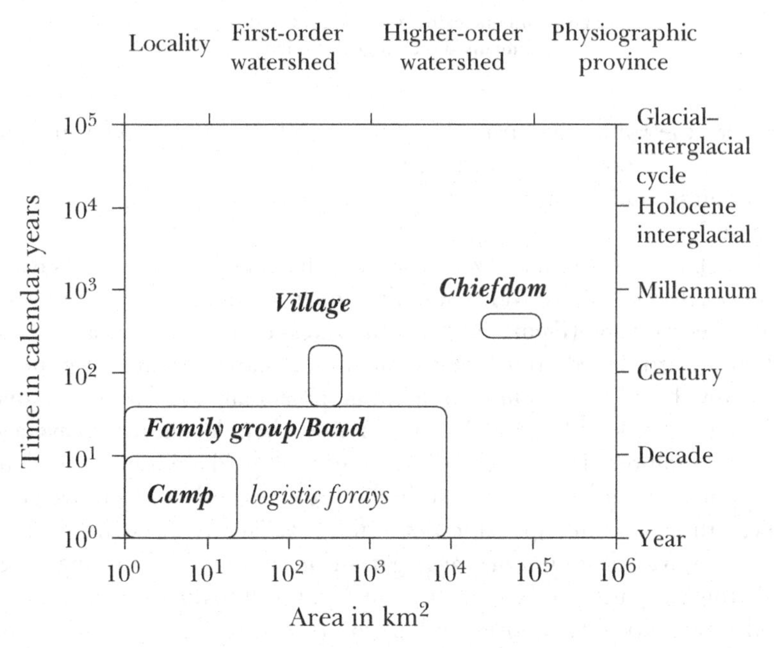

Figure 3.3 Spatial and temporal extent of ecological impacts of prehistoric human populations (adapted from Johnson and Earle, 2000).

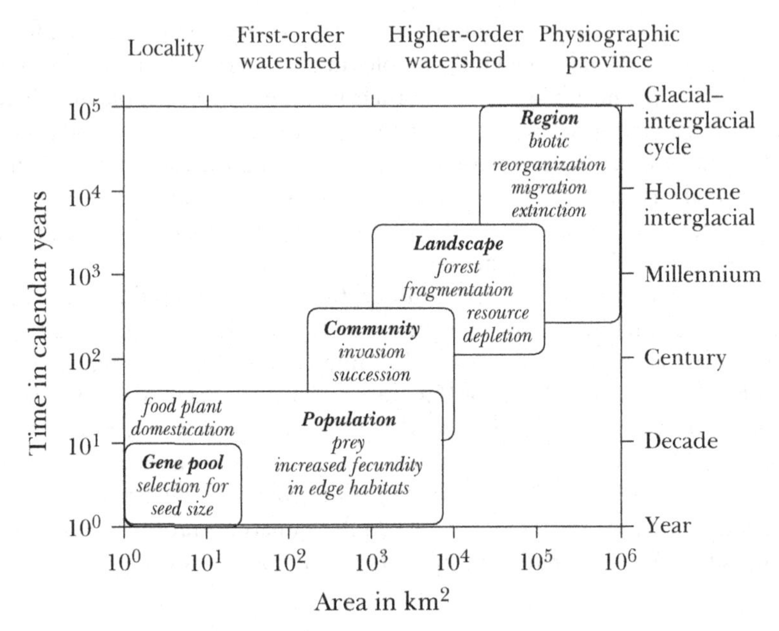

Figure 3.4 Spatial and temporal extent of ecological feedbacks between prehistoric humans and their environment.

provides the habitat spectrum for gathering of wild plant foods and for collecting of wood for hearth fires and habitations. A much larger search radius on the order of tens of kilometers is utilized for hunting and fishing (Figure 3.4).

This basic panarchical level represents the foraging mode of subsistence (Kelly, 1995) by which mobile human populations interact with natural ecosystems (Figure 3.5). As characterized by Johnson and Earle (2000), hearth-based groups of several families have low rates of population growth and low population densities, potentially much less than one person per 100 km^2 (Figures 3.2, 3.3). Rather than depending on average overall conditions, foragers with limited technological abilities must adapt to population-regulating shortfalls in food reserves and must survive periodic, although unanticipated, crises in food availability or environmental variability. Mobile groups may aggregate during autumn for hunting or for gathering ripe nuts, yet they can disperse in response to fluctuating or sparse food availability. The relatively unlimited movement of such groups across vaguely defined home territories minimizes competition among groups for resources.

As human densities increase through time, and the landscape is carved into mutually bounded territories, population mobility becomes restricted to locally available resources. Foraging efficiency for household subsistence needs requires more intensive use of existing food sources, which may lead to depletion of favored items and switching to less desirable or less nutritious alternatives. This intensified resource competition may heighten the territorial imperative to build group alliances, not only as defensive solutions to protect access to productive habitats, but also for greater communal investment in technology to ensure ample food production, and for developing expanded trade networks. These larger, integrated groups select leaders at first to guide, then later to manage, their activities.

A second panarchical level of human–ecosystem interaction is shaped by the social organization of village groups of forager–horticulturalists (Figure 3.5). Johnson and Earle (2000) considered the political integration of autonomous local groups, tethered to a local village, as the cultural manifestation of mutual needs to augment food production, to defend homelands from organized raiding and warfare, and to enhance opportunities for mate selection. Such village-sized groups consist of several hundred people occupying territories ranging from 300 to 500 km^2 (Figures 3.2, 3.3), with population densities approaching one person per square kilometer of land. These groups would supplement foraging with limited-scale horticulture in order to provide predictable harvests of storable plant food. Establishment of more sedentary groups would encourage use of a broader spectrum of native plant species, and intentional seed selection leading to plant domestication (Figure 3.4). Local groups might stay at a given village site for between 50 and 300 years, depending upon their rate of consumption of nearby firewood reserves and diminishing food production tied to depletion of soil nutrients within garden plots. Cultural impacts would trigger plant community responses, including invasion by species of weeds, and secondary succession of woody plants on the patchwork of abandoned horticultural plots (Figure 3.4). Communal activities for preparing new plots to replace fallow fields, and the growing-season round of sowing seed, weeding, harvesting, processing, and storing crops, all are examples of cooperative efforts aimed at sustaining food production levels. By developing regional trade networks, individual villages would be buffered from sporadic crop failures, and overall human population size would increase. Social transformation of these "neighborhood societies" (Johnson and Earle, 2000) leads to the domestication of plants and animals, as sedentary human village groups adopt an ownership sense of land tenure. Both ancestor-granted rights as property stakeholder

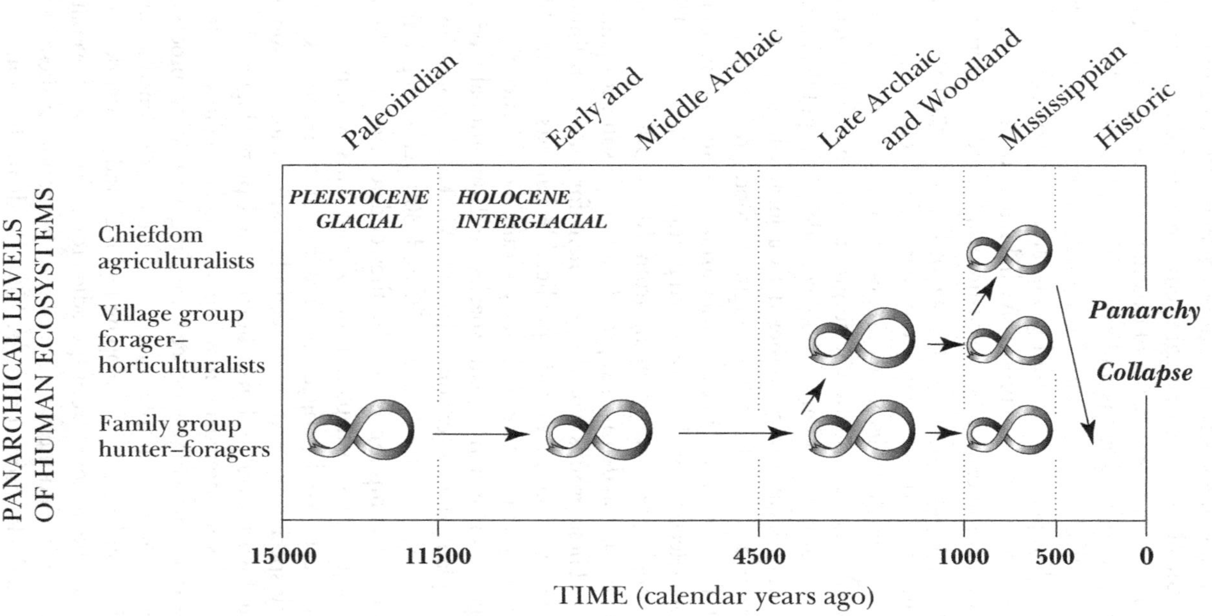

Figure 3.5 Development of panarchical levels of human ecosystems in North America during the past 15 000 calendar years.

and ceremonial rituals bind forager–horticulturalists within their ecological neighborhoods.

A third panarchical level of human–ecosystem interaction develops when agrarian societies are capable of transforming native ecosystems at landscape and regional scales (Figure 3.5). This corresponds with the cultural evolution of regional polities, political institutions that organize human groups into chiefdoms, and broader-scale nation states (Johnson and Earle, 2000). As local groups are interlinked and governed by more comprehensive institutions, social stratification leads to the emergence of a powerful elite class who maintain control over production and distribution of agricultural commodities, trade of high-status items, accumulation of material wealth, ideological direction of ceremonial rituals defining societal norms, and military dominion and conquest of new territories.

The highest social organization attained in prehistoric times in eastern North America was the development of chiefdoms between approximately 1000 and 500 BP. Populations of these chiefdoms ranged from thousands up to tens of thousands, and defended territories were as large as 10^5 km^2 (Hudson, 1997). High densities of farmers (more than 50 individuals supported per square kilometer of arable land) were concentrated along the fertile riparian corridors of floodplain and low-lying terraces. Individual chiefdoms experienced social consolidation and conquest. The rise, eclipse, and abandonment of individual chiefdoms were periodic, occurring over time spans of 300 to 600 years (Smith, 1978; Anderson, 1994; Milner, 1998) (Figures 3.2, 3.3). During the existence of the panarchical level of human ecosystem represented by chiefdoms, concentrated human land use generated a landscape patchwork of fragmented forests, cultivated land, and nutrient-depleted old fields abandoned as fallow land (Figure 3.4).

Johnson and Earle (2000) characterized the prehistoric emergence of even more administratively complex and regionally organized societies elsewhere in the New World during the past millennium. These agrarian nation states and empires were centrally controlled by ruling bureaucracies that managed both the production of crops and the redistribution of food reserves, traded as commodities by a merchant class. Ethnically diverse people numbered from the hundreds of thousands up to several million. For example, in Central America the prehistoric Maya State held a population base estimated between 3 and 14 million, and a territory of 1.6×10^5 km^2 in a regional state dominant for 300 years (from 1400 to 1100 BP) (Binford *et al.*, 1987). In the Andes of westernmost South America, the Inka Empire represented a populace between 8 and 14 million individuals, supported by irrigated agriculture, fishing, and animal husbandry, and a domain of 9×10^5 km^2. Within these socially stratified societies, a ruling elite governed by military might, and a merchant class maintained economic control over rural agriculturalists (Johnson and Earle, 2000).

PALEO-PANARCHY OF THE EASTERN WOODLANDS

Prehistoric Native Americans superimposed a cultural overprint that shaped ecosystem dynamics by a patchwork pattern of immigration and settlement, by competition for scarce or valued resources, and by geographic displacement of populations in the quest for territorial dominance (Bonnischen and Turnmire, 1999a,b). This prehistoric cultural legacy has dictated where, when, and how pervasively human interactions and management practices were embedded within their environment.

Founding groups of the first Paleoindian hunter–foragers came from as many as five discrete source populations (Meltzer, 1993; Forster *et al.*, 1996; Brown *et al.*, 1998; Schurr and Wallace, 1999). Small family bands immigrated across the Siberian–Alaskan land bridge of Beringia, along the coastline of British Columbia (Fladmark, 1979, 1990), and along ice shelves and pack-ice margins of the northern Pacific Ocean (Dixon, 1999) and northern Atlantic Ocean (Stanford and Bradley, 2000). Archaeological evidence for preClovis PaleoAmericans is enigmatic and ambiguous, interpreted as "failed migrations" for populations generally below the detectable threshold of archaeological visibility (Anderson, 2001; Anderson and Gillam, 2001). With only Old World immunities to "heirloom" pathogens, the earliest immigrants may have been vulnerable to the population-winnowing exposure to endemic, New World diseases (Dillehay, 1991).

Compelling and undisputed archaeological evidence, dated from 13 500 to 12 900 BP, documents the geographic radiation of nomadic Clovis hunter–foragers across the Americas. During that time interval, Paleoindians dispersed preferentially along major river corridors and coastal routes (Anderson and Gillam, 2000). Named for the archaeological site near Clovis, New Mexico, where their distinctive projectile points were first recovered with remains of now-extinct mammoth, Clovis hunters employed a more effective weapon technology than their predecessors; they used fluted stone spear points to hunt large game (Dixon, 1999). Clovis invaders were part of a destabilizing "pulse phenomenon" (Bonnischen *et al.*, 1987). These hunters of Pleistocene megafauna arrived during the time span characterized by greatest magnitude and rate of postglacial warming, and by wholesale biotic reorganization.

The panarchical restructuring of late-glacial ecosystems 13 000 BP coincided with massive climate change and glacial retreat, assembly of ephemeral plant communities with poor modern-day analogs, and extinction of many genera of Pleistocene megafauna (Lambert and Holling, 1998). The concomitant entry of a superior human predator displaced faunal carnivores as the dominant keystone species regulating trophic dynamics (Whitney-Smith, 1998). Late-glacial humans may have destabilized predator–prey dynamics as prehistoric ecological imperialists (Crosby, 1986). They also possibly served as a critical vector for introducing one or more contagious hyper-diseases that may have decimated megafauna herbivores (MacPhee and Marx, 1997). In this way, vital roles of native browsers and grazers as landscape-disturbance generators and seed dispersers may have been terminated (Janzen and Martin, 1982).

With the onset of Holocene interglacial conditions, nomadic family-group bands expanded from initial staging areas, and settled more intensively across regional watersheds (Figure 3.2), drawn to habitats rich in small game, a diverse suite of plant foods, and chert quarry sites for stone tools (Anderson, 2001). During the Archaic cultural period (11 500 to 3200 BP), accelerated human population growth is inferred from increased site frequency (measured as numbers of archaeological sites per consecutive intervals of 100 years). Bands of nuclear families of hunter–foragers established an annual round of activities. They moved seasonally along riverways among a network of strategically located base camps, and they dispersed radially on logistically organized foraging trips to favored resource patches. As geographically distinct groups adapted to local settings, cultural divergence in weapon manufacture resulted in stylistic differences of stone projectiles. These projectiles were shaped and reduced in size to be more appropriate for hunting the species of small game animals that survived the late Pleistocene extinction event. Archaic people responded opportunistically during the early Holocene to changes in the composition and extent of temperate deciduous forests. Oak, hickory, chestnut, and walnut became widespread and provided a bountiful autumn harvest for foragers, who collected, heat-treated by parching, and cached the mast for winter food reserves (Anderson, 2001).

During the Late Archaic (4500 to 3200 BP) and Woodland (3200 to 1000 BP) cultural periods, human populations continued to increase, and people subdivided the available landscape into progressively smaller, mutually bounded territories. Autonomous local groups, with hundreds of people forming several kinship clans, established more permanent villages and village groups (Figure 3.2) organized around communal food storage or defense (Johnson and Earle, 2000). Rather than shifting along riverways, the seasonal round was centrally based at the floodplain village sites, termed "domestilocalities" by Smith (1992). Individuals on logistical forays moved cross-valley into the adjacent uplands. Cultural adaptations optimized both foraging and food production via incipient horticulture of native weedy plants in garden plots, increasing the number of panarchical levels from one to two (Figure 3.5). After 3200 BP, technological innovations in manufacture of ceramic pottery created practical vessels for storing food and seed grains from one growing season to the next.

During the Mississippian cultural period (1000 to 500 BP), organization of regional polities – in North America, chiefdoms involving many thousands of individuals and in Central and South America, nation-states involving millions of people – developed as socially ranked, religious and political systems (Figure 3.2), supported by agricultural ecosystems of the exotic cultivars, maize, squash, and beans (Anderson, 2001). Governed by elite agrarian societies, territorial expansion by conquest and economic control by trade network placed high premium upon dependable crop harvest. Social infrastructure required arable lands suitable for intensive agriculture and the farming populations necessary to generate crop surpluses for barter, toll, or insurance against future poor harvests. During the Mississippian cultural period, three levels of the human ecosystem panarchy were maintained simultaneously (Figure 3.5) as a cultural mosaic across the Eastern Woodlands.

With historic contact and conquest by European Americans in the past 500 calendar years, ecological imperialism by warfare, introduction of domesticated herds and crops, and spread of Old World disease triggered a collapse of the panarchy of traditional Native American ecosystems (Figure 3.5). Subsequent reorganization has occurred within the context of an industrialized, globally interlinked society (Johnson and Earle 2000).

CONCLUSIONS: HUMAN-MANAGED ECOSYSTEMS

The transformation of natural ecological systems to human-managed ecosystems makes the Holocene unique among interglacial intervals (Delcourt, 2002). We propose that new panarchical levels in prehistoric Native American ecosystems emerged during the late Quaternary, with humans adopting new lifeways through adaptive interactions with their biotic and abiotic environment (Figure 3.1). Through the Holocene interglacial, human population growth has tracked changing behavioral strategies and technological innovations, allowing for increases in quantities, diversity, and reliability of the subsistence base as people found new ways to procure, process, and store food reserves. Three distinct panarchy levels exhibit qualitatively different modes of traditional Native American–ecosystem interaction, shifting in balance through time and across space from predominantly natural adaptive cycles to ones increasingly interlinked with anthropogenic activities.

PART II

Ecological feedbacks and processes

OVERVIEW

Complex ecological feedbacks between humans and their environment have taken place on many different spatial and temporal scales (Figure 3.4). To look at cycles of adaptation, organization, destabilization, and reorganization in their appropriate contexts (Figure 3.1) requires being able to slide back and forth in time and place, as relevant to the panarchical level of interest. In Part II, we summarize examples of prehistoric Native American ecosystem dynamics based on level of biological interaction, from genetic and population level to community, landscape, and regional level.

In Chapter 4, *Gene-level interactions*, we explore the ecological implications of the transition from foragers to farmers that occurred beginning in the Archaic cultural period. This transition in lifeways resulted in the addition of a panarchical level (Figure 3.5), and was characterized by a number of facilitative interactions among plants, their pollinators, their seed dispersers, and their harvesters. Incipient domestication of native plant species began by 4500 BP. In the process of becoming farmers, Archaic Native Americans developed mutualistic relationships with individual plant domesticates as they selected phenotypes of edible plants that showed the greatest robustness of seeds and fruits. With first domestication, genetic diversity was enhanced locally for squash, sunflower, maygrass, little barley, goosefoot, and marsh elder.

Chapter 5, *Population-level interactions*, explores the relationship between Archaic and Woodland people and their plant domesticates, as well as their animal prey. Archaic people utilized mast nuts as a major part of their diet and may have protected groves of masting trees, in addition to advancing the northward spread of tree species including hickory, walnut, and chestnut during the early and mid-Holocene. During the Archaic and Woodland cultural periods, populations of small game, such as white-tailed deer, were affected by the seasonal round of human activities across the territories of hunter–forager groups. Human disturbance of vegetation surrounding village sites would have increased forest edge would have favored white-tailed deer population growth, offset by seasonal hunting. Development of semi-permanent residential sites along riparian corridors would have opened up light gaps suitable for colonization and spread of weedy ruderal plants, whose seeds were then harvested during the time of plant domestication.

Examples cited in Chapter 6, *Community-level interactions*, demonstrate how prehistoric Native Americans had widespread effects on biotic communities. Late-Archaic and Woodland people were forager–horticulturalists who concentrated their farming activities not only in river bottoms but also near cliff-edge rock shelters from the Ozark Plateaus to the southern Appalachian Mountains. The strong coincidence of prehistoric human occupation, including ethnobotanical remains indicating domestication of native plants, with fossil charcoal documenting increases in local fires presents a convincing case for a cause-and-effect relationship between Native American activities and changes in forest vegetation on the Cumberland Plateau of Kentucky during the past 4500 years. The cumulative impact of human activities resulted in a fine-grained patchwork of vegetation that included fire-adapted pine forest on ridgetops, fire-tolerant oak–chestnut forest on upper slopes, and fire-intolerant mixed mesophytic forest on mid- and lower slopes. Prehistoric Native Americans generated an intermediate disturbance regime that increased biological diversity both within and between plant communities.

In Chapter 7, *Landscape-level interactions*, we define critical thresholds for landscape change in relation to prehistoric human activities (Figure 3.4). The case study from Crawford Lake, southern Ontario, illustrates how even small groups of sedentary people can have cumulative impacts on landscapes surrounding village sites. In the Mississippian cultural heartland, with development of the third panarchical level of chiefdoms, major landscape modifications occurred. Over-exploitation of resources led to forest fragmentation, soil degradation, and deterioration of the environment across broad landscapes. In the Little Tennessee River Valley of eastern Tennessee, detailed archaeological and paleoecological records show that human settlement resulted in expansion of agricultural ecosystems and changes in environments during Woodland and Mississippian times. Together, these studies show that when people became tethered to fixed sites for more than two centuries, and when population densities exceeded 50 people per km^2 of cultivated land, cultural activities resulted in environmental degradation as native ecosystems became fragmented and discontinuous.

On the American Bottom of southern Illinois in the Central Mississippi River valley, about 1000 BP, settlements ranging from hamlets to the chiefdom of Cahokia occupied an extensive area across most of the levee crest habitats in the active river meander train. Forests that in the Woodland cultural period occupied these bottomland habitats were highly fragmented by the Mississippian cultural period. At the prehistoric metropolis of Cahokia, a critical ecological threshold was surpassed when climatic and geomorphic changes destabilized the rivers along which prehistoric people were concentrated, a major factor leading to collapse of the Native American ecosystem panarchy and abandonment of the Mississippian heartland by 600 BP.

Chapter 8, *Regional-level interactions*, explores the changes in human ecosystems during the major reorganization of macro-scale adaptive cycles at the transition from the late Pleistocene to the Holocene (Figure 3.5). During the late-glacial interval, ecosystems were poised for radical change. Arriving during the changeover from glacial to interglacial climate, Clovis hunters and their Paleoindian successors selectively harvested native big-game animals, displacing Pleistocene carnivores as keystone predators. Disruption of trophic-level dynamics contributed to extinctions, vacated niches, and the precipitous disassembly of ice-age megafaunal communities at the end of the Pleistocene.

4
Gene-level interactions

FROM FORAGING TO CULTIVATION TO PLANT DOMESTICATION

Native Americans have used plants for food, fiber, dyes, medicine, shelter, and fuel throughout their history on the North American continent. The use of plants changed over time as shifts in biotic resources resulted in new possibilities for exploitation of plants. Native Americans were gardeners on a local scale beginning as early as 7800 BP; the development of extensive farming did not take place until about 950 BP with expansion of maize agriculture (Fritz, 1990). The origin of agriculture in the New World involved the interplay of climate change and human–plant mutualistic interactions resulting in genetic selection for productive plant phenotypes.

A continuum of human–plant interactions (Figure 4.1) leads from foraging, through cultivation, to the domestication of plants (Ford, 1979, 1985). Initially, harvesting edible plants may have had minimal genetic impact on the plant populations. Through inadvertent dispersal of seeds and trampling of inedible plant competitors, however, certain food plants may have been encouraged. Later they may have been tended by deliberate weeding out of undesired plants, by dividing and transplanting root clumps, or by sowing of seeds. Cultivation may take the form of tillage, resulting from digging corms, bulbs, tubers, and rhizomes (Anderson, 1997). Cultivation may lead to morphological modification, particularly of fruits and seeds, through deliberate selection and storage, in which case the plant is referred to as domesticated (Ford, 1981; Fritz, 1990).

Gradient of increased work effort

Harris (1989) characterized human–plant interactions as a gradient of increasing work effort (human energy expended per unit area of exploited land) related to (1) increasing size of human populations; (2) aggregation into permanent settlements; (3) a changing in seasonal schedule to tend

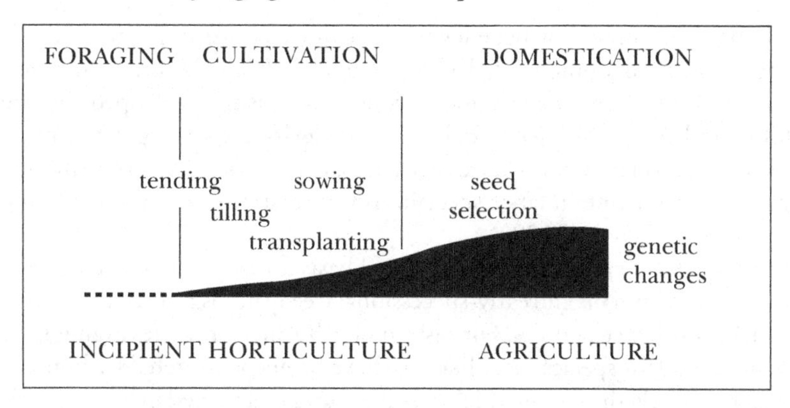

Figure 4.1 Evolutionary continuum of human–plant interactions associated with the harvesting of plant food (modified from Ford, 1985).

fields; and (4) division of labor based upon gender or age. Broad-scale foraging for wild plants requires less work effort than does the deliberate production of plant food. Protection or planting of prized resource patches such as nut groves or floodplain weeds requires crossing an energy investment threshold and represents the first steps toward domestication (Harris, 1989; Rindos, 1989). Limited forest clearance creates disturbed patches for planting small garden plots. Soil tilling, broadcast sowing of seed, and transplanting of propagules to newly opened habitats involve returning to fields at least periodically to tend growing crops. Harvesting of plant foods requires both rudimentary processing such as parching of seed as well as effective storage of foodstuffs.

Crossing a second energy threshold from wild plant propagation to active cultivation is an ecologically significant transition (Harris, 1989). Cultivation requires regular land clearance, systematic tillage of soil, and planting. The third energy threshold proposed by Harris (1989) involves deliberate selection of plant variants, leading to domesticated taxa dependent upon continued propagation by humans. The cultural transition to maintaining agricultural ecosystems links farmers to ongoing obligations of refurbishing soil fertility, weeding, selection and storage of the next year's seed stock, and defending the harvest against potential predators attracted to the cultivated fields.

Human–plant mutualisms

Evolutionary ecological models favor the view of agriculture as a mutualistic process (Crites, 1987), reflecting long-term human behavioral modification with positive, mutually beneficial interactions between populations of people and their domesticated plants (Rindos, 1984; Winterhalder and Goland, 1997). In part, human–plant mutualisms developed in the Holocene because increasing cultural complexity required greater, more predictable food reserves as barter in exchange networks, to gain ceremonial status, and to maintain labor specialization and stratification of emerging social classes (Hayden, 1992).

Crites (1987) extended the conceptual basis for the mutualistic interaction between humans and early-successional weedy herbs favored by regular, annual disturbance regimes. For eastern North America, Crites highlighted the suite of plant species most likely to have been encountered by hunter–foragers, namely the plants widely dispersed and established in "home territories" defined by human habitation and their foraging range. Convincing ethnobotanical evidence from Middle Tennessee shows that morphologic characteristics of several species of cultivated plants diverged from their natural wild populations during the interval from about 2200 to 1250 BP in the Woodland cultural period. Seeds of wild plants stored as food reserves substantiate the primary dependence of humans upon the harvest of cultivated native weeds such as reticulate-seeded goosefoot (*Chenopodium bushianum/berlandieri* and *C. missouriense*), maygrass (*Phalaris caroliniana*), erect knotweed (*Polygonum erectum*), and domesticated native species of sunflower (*Helianthus annuus*), marsh elder or sumpweed (*Iva annua*), squash (*Cucurbita pepo*), and maize (*Zea mays*). Each of these plant taxa was represented in all the ethnobotanical samples from four cultural intervals of Middle Woodland age. The reliance upon domesticated plants as food resources coincides with archaeological evidence for the development of year-round communities. The study by Crites (1987) illustrates that the niche breadth and gene pool of edible plants broadened as a direct consequence of human–plant mutualism during the domestication process.

MULTIPLE PATHWAYS TO PLANT DOMESTICATION IN EASTERN NORTH AMERICA

Early concepts of the introduction and spread of exotic cultigens, as well as of the relative roles of native plant species versus exotics in the prehistoric Native American diet (Yarnell, 1976, 1977; Ford, 1981) have been substantially revised based upon (1) extensive analysis of carbonized ethnobotanical remains screened from hearths and other archaeological contexts, separated by water flotation (Struever, 1968; Chapman and Watson, 1993);

(2) temporal evidence from Accelerator Mass Spectometry (AMS) radiocarbon dates of individual seeds, fruits, and rinds (Decker-Walters, 1993; Fritz, 1995); (3) studies of seed morphology (Smith, 1989; Decker and Wilson, 1986; Gremillion, 1993a); and (4) allozyme and other evidence elucidating the evolutionary relationships among taxa of cultivated plants (Hurd *et al.*, 1971; Kaplan, 1981; Decker-Walters *et al.*, 1993; Fritz, 1990, 1999).

The traditional "triumvirate" of maize (*Zea mays*), squash (*Cucurbita pepo*), and beans (*Phaseolis vulgaris*) that were grown together in mixed agriculture in eastern North America at the time of European contact was an agricultural system that had been in practice there for only a few hundred years (Fritz, 1990). The history of introduction and domestication was different for each species. Ethnobotanical evidence indicates that while hunter–gatherers first used squash and gourd beginning 7800 BP, agricultural societies in which maize, squash, and beans played important nutritional roles did not develop until 950 BP. In the intervening millennia, the paths toward plant domestication followed different trajectories in different subregions of eastern North America (Fritz, 1990).

The first widely cultivated native plants were those with starchy and oily seeds, known from archaeological sites dating generally to 3000 BP in the Ozarks, the Midwest, and the Midsouth. In the Midwest, in the American Bottom of the Illinois River Valley, the food-production shift to agriculture in the early Mississippian period was based primarily on maize (10- to 12-row variety), without beans. In the central Ohio River Valley (McLauchlan, 2003), Woodland archaeological sites contain fewer remains of native species of food plants than those farther south and west; sites representing the Mississippian-age Fort Ancient culture indicate that cultivation of maize (8-row variety) began together with beans. To the south of northern Alabama, Woodland period people did not grow native seed crops extensively; later, in the early Mississippian, they grew maize (other than eastern 8-row variety) without beans. In the Lower Mississippi Valley, the Eastern Agricultural Complex was not as important as in the Midwest, but squash and gourd were represented before 3200 BP. To the west, in the Ozark Highlands of eastern Oklahoma, plant cultivation generally followed the Midwestern pattern, with beans uncommon until late Mississippian times. A southwestern influence was reflected in the use of amaranth (*Amaranthus*) and cushaw squash (*Cucurbita argyrosperma*), neither of which was cultivated to the east. In the northeastern United States, 8-row maize and beans were introduced relatively late, about 950 to 650 BP; there, some but not all foragers became agriculturalists after that time (Fritz, 1990).

THE EASTERN AGRICULTURAL COMPLEX

A mechanism by which first plant domestication took place in the mid-continental United States (Figure 4.2) was proposed by Smith (1987, 1989, 1992). He suggested that during the mid-Holocene (about 8900 to

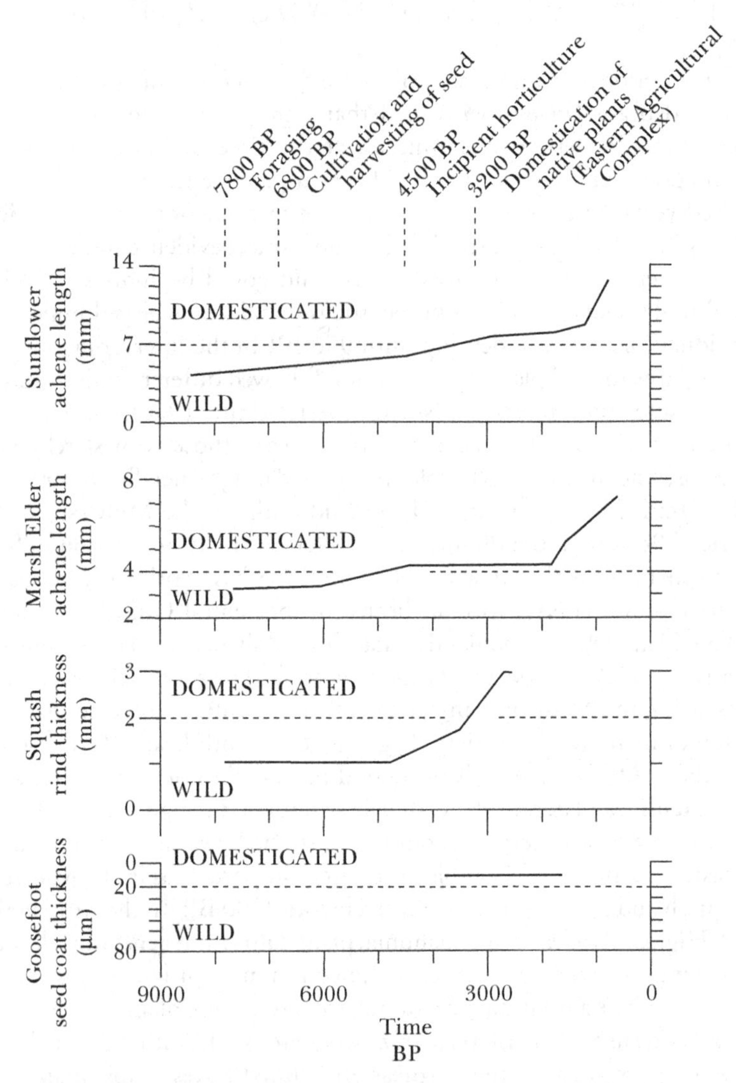

Figure 4.2 Coevolutionary interactions of humans and weedy plant species within riparian habitats, with "domestilocalities" serving as settings for morphologic selection, cultivation, and domestication of native plants between 4500 and 3200 BP (modified from Smith, 1992).

4500 BP), climate warming led to changes in stream flow as rivers such as the Mississippi changed from braided to meandering regimes, and floodplain vegetation stabilized along riparian corridors. With an increase in aquatic productivity accompanying shoaling of streams throughout the Midwest, people were drawn to inhabit floodplain sites in localized "domestilocalities," which became the first settings for plant cultivation. Riparian corridors were ecologically favorable for colonization by weedy ruderal plants because of high soil nutrients and anthropogenic disturbance that locally opened up light gaps into which seeds could readily disperse. In these environments, r-selected plants thrived, increasing their production of seeds and decreasing the length of seed dormancy. Such microevolutionary changes would have taken place without direct human selection, but they would have been encouraged indirectly by anthropogenic disturbance that kept the riparian habitat open.

Once people began to use plants, however, they selectively began to encourage the growth of economically desirable species, first with weeding to reduce interspecific competition, then planting and reharvesting seed. This would have selected for increased retention of seeds and more compact, dense seed heads with seeds maturing synchronously, all of which led to increased yield of harvestable plant food. Seedbed competition would have led to reduced dormancy (expressed, as in goosefoot, by thinner seed coats), more rapid growth, and increased seed size. The morphological changes observed in squash and gourd, marsh elder, goosefoot, and sunflower in the archaeological record dating from 4500 to 3200 BP (Figure 4.2) are considered to be the product of prehistoric plant domestication (Smith, 1987, 1989, 1992).

Application of foraging theory to domestication of plants in eastern North America was foreshadowed by the early work of Anderson (1952). His "dump-heap" theory of plant domestication maintained that dooryard gardens were sites for initial domestication of New World plants, because of anthropogenic disturbance that enriched soils on refuse heaps and opened up well-lighted openings in the forest canopy suitable for colonization by weedy adventive species.

By 5700 BP, native species of marsh elder and goosefoot were being utilized by prehistoric people in west-central Illinois. Between 4500 and 3200 BP many starchy and oily seeds of plants native to eastern North America were being domesticated, including not only marsh elder and goosefoot (Smith, 1992), but also sunflower (Crites, 1993), maygrass, and little barley (*Hordeum pusillum*) (Yarnell, 1976, 1977; Watson, 1989). By 2600 BP a "full-fledged" horticultural complex was being utilized in the mid-continent region extending from Arkansas to Kentucky and Tennessee.

Variously called the "Eastern Agricultural Complex" (Smith, 1992) and the "early Woodland garden complex" (Watson, 1989), the plant species involved included two container species (*Cucurbita pepo* and *Lagenaria siceraria*), two oily-seeded species (*Iva annuua* and *Helianthus annuus*), and two starchy-seeded species (*Chenopodium berlandieri* and *Phalaris caroliniensis*). Other starchy-seeded species added later included erect knotwood (*Polygonum erectum*), little barley, and giant ragweed (*Ambrosia trifida*). Middle Woodland gardens also included tobacco (*Nicotiana rustica*). By late-prehistoric times, "a panoply of tolerated, encouraged, or quasi-cultivated plants" were grown (Watson, 1989), with as many as 20 additional species used for their edible seeds and fruits (amaranth, *Amaranthus* sp.; maypops, *Passiflora incarnata*; wild beans, *Strophostyles helvola* and *Phaseolus polystachios*; ground nut, *Apios americana*), as potherbs (purslane, *Portulaca oleracea*; pokeweed, *Phytolacca americana*; Jerusalem artichoke, *Helianthus tuberosus*; spurge, *Euphorbia maculata*; carpetweed, *Mollugo verticillata*), or as fiber and dye plants (Indian hemp, *Apocynum cannabinum*; milkweed, *Asclepias tuberosa*; rattlesnake master, *Eryngium yuccifolium;* nettles, *Laportea* and *Urtica dioica*; mulberry, *Morus rubra*; staghorn sumac, *Rhus*; bedstraw, *Galium*), for extracting salt (water lotus, *Nelumbo*) and for medicinal, hallucinogenic, and ceremonial drinks (black nightshade, *Solanum americanum*; yaupon, *Ilex vomitoria*) (Yarnell, 1976, 1977; Watson, 1989; Drooker, 1992, Jakes and Sibley, 1994; Andrews and Adovasio, 1996; Kuttruff *et al.*, 1998).

CUCURBITA PEPO: LASTING LEGACY OF HUMAN–PLANT INTERACTIONS

Paleoethnobotanical record from North America

Most known wild species of the genus *Cucurbita* (squashes and gourds) are restricted to the region from Mexico to the southwestern United States. In eastern North America, the earliest record of indigenous plants that later became widely domesticated is of pepo gourd *(Cucurbita pepo)* associated with mastodon remains from the Aucilla River in Florida, dated to 14 700 BP (Newsom *et al.*, 1993). Bottle gourd (*Lagenaria siceraria*) was first known in the southeastern United States at Windover, Florida, dating to 7800 BP (Doran *et al.*, 1990). These first occurrences of what later became domesticates were not demonstrably introduced by humans from MesoAmerica. Rather, the southeastern United States was an independent center of origin for *Cucurbita*, and *Lagenaria* gourds may have floated in on ocean currents from the Caribbean (Fritz, 1990; Heiser, 1989). In the mid- to late Holocene, many varieties of *Cucurbita* were developed as domesticates throughout the Americas (Hurd *et al.*, 1971).

Cucurbita remains, from archaeological sites in Tennessee and Illinois, date to 7800 BP (Crites, 1991). A *Cucurbita pepo* rind fragment from the Sharrow site in Maine dates to 6500 BP (Peterson and Asch Sidell, 1996). Squash seeds from rockshelter sites in eastern Kentucky, dating from 5900 to 2900 BP (Cowan, 1997), and those recovered from Phillips Spring, Missouri, dating from 4850 BP, are evidence for early domestication of squash (Kay *et al.*, 1980). By 4500 BP, both pepo and bottle gourds were in general use in eastern North America. In eastern Kentucky, after 2900 BP, new types of container gourds were developed, and by 1300 BP, edible fleshy forms of squash were domesticated (Cowan, 1997).

Based on allozyme evidence, Decker-Walters (1993) and Decker-Walters *et al.* (1993) determined that *Cucurbita pepo* includes three distinct taxa with different distributional histories. *Cucurbita pepo* subsp. *ovifera* var. *ovifera* consists of cultivars of squashes similar to modern yellow crooknecks, pattypans or scallops, and small ornamental gourds. These cultivated forms were developed in eastern North America from wild progenitors. A second subspecies, *pepo*, which today includes zucchinis and pumpkins, originated in Mexico. Wild *C. pepo* var. *texana* from Texas, Alabama, Arkansas, and Illinois, are allozymically very similar to eastern North American cultivars. Variety *texana* is thought to be the progenitor of var. *ovifera* and probably had a more widespread distribution in eastern North America in prehistoric times than at present. Seeds from preColumbian contexts on Hontoon Island, Florida, are morphologically similar to *C. pepo* var. *texana*. Subspecies *ovifera* and *pepo* thus have had distinct temporal–spatial histories in ancient North America. These data demonstrate independent cultivation from divergent populations of the original wild species in North America and Mexico.

Invasional meltdown: positive interactions between cucurbits and bees

The process of domestication of squash/gourd in North America was facilitated by the coevolutionary adaptations between *Cucurbita* and its pollinator bees (*Peponapis* and *Xenoglossa*), which are adapted to collect the large (80 to 150 μm diameter) and spiny pollen grains, and to drink the nectar of *Cucurbita*, from which the bees derive most of their food (Hurd *et al.*, 1971).

Hurd *et al.* (1971) suggested that the role of humans was to have discovered the species of *Cucurbita* already evolved in the flora of Mexico, to have used them opportunistically, later to have domesticated and cultivated them, then to have transported and grown them well beyond their native ranges. Hurd *et al.* (1971) speculated that the original ranges of the bees were altered following the spread of domesticated cucurbits, as the bees extended their ranges using "pollen avenues" established by cultivated cucurbits. The role of the bees is thus a coevolutionary facilitation with the cucurbits.

The prehistoric spread of cucurbits and their bees throughout much of temperate North America is a possible example of "invasional meltdown" (Simberloff and Von Holle, 1999) – a positive interaction between species of previously restricted ranges that became more widespread because of human intervention. In the southeastern United States, expansion in the range of *C. pepo* var. *texana* may have occurred as populations from Texas were spread eastward along the Gulf Coastal Plain, becoming naturalized along riverine corridors, where Archaic people used a seasonal pattern of resource exploitation that was aligned with watershed distributions (Anderson and Sassaman, 1996). Native cucurbits were wild, weedy plants that were aggressive and successful colonizers that would have flourished on soils disturbed by human activities (Smith, 1987). Early squash and gourd remains from Middle Archaic sites in Missouri and Illinois may have been wild types that spread along human-disturbed riparian corridors, but may not have been cultivated until after 4500 BP. After that time, seeds and rinds from archaeological contexts show increases in size and thickness that indicate human selection for edible varieties of squash (Figure 4.2).

Mid-Holocene gourds from the Midwest and New England

Did the initial use and cultivation of gourds trigger a transition to gardening and incipient farming in eastern North America? Or was it instead an integral part of a fisher–gatherer–hunter lifeway? The location in Maine where *Cucurbita* was found in archaeological context dating to 6500 BP is today north of the limit of cultivation of gourd, and Fritz (1999) speculated that people were responsible for spreading *Cucurbita pepo* in cultivation beyond the range it would have occupied naturally. She further speculated that it may have been used for fishnet floats, bobbers, and small containers rather than for food.

Asch (1994) suggested that Archaic hunter–gatherers planted and tended *Cucurbita pepo* and were responsible for its range extension in the mid-Holocene from its indigenous center in the southeastern United States, passing the gourds upstream along trade networks to the Midwest as well as along the Atlantic Seaboard. Most Archaic sites are located in river valleys where fish were an important part of the diet, and sites where cordage is well preserved indicate that fishnets were in use, making the hypothesis of use of cucurbits as bobbers and fishnet floats plausible. People probably also ate cucurbit seeds (Cowan, 1997) and flowers (Gremillion and Sobolik, 1996), and the gourds may have been used as rattles (Fritz, 1999). Small-sized cucurbit gourds dating to 7800 BP from archaeological sites in the lower Midwest thus may have been cultivated, but people initially may have deliberately selected for small-sized gourds used in fishing, rather than being interested in developing large, fleshy fruits and large edible seeds, as were characteristic of later domesticated varieties. Fritz (1999, pp. 426–7) proposed that "people passed gourds northward from their native habitat on the Gulf Coastal Plain as part of a fishing technology that spread as river systems stabilized. Although gourds were easy to grow, they required tending and periodic replanting. These activities were fully compatible with a hunter–gatherer lifeway and did not trigger an agricultural revolution."

THE ORIGIN OF THE COMMON DOMESTICATED BEAN

The first archaeological record of the common bean (*Phaseolis vulgaris*) is from Coxcatlan Cave, Tehuacan Valley, Mexico, dating from 7800 to 6300 BP. Bean cultivation became important and widespread between 1700 and 1150 BP. The common domesticated bean (*Phaseolis vulgaris*) was not introduced to eastern North America until about 650 BP (Hart and Scarry, 1999).

Today, wild bean types grow in isolated stands in a long and disjunct range from west-central Mexico through Central America, along the eastern slopes of the Andes, to Argentina. One group evolved in the Mexico–Guatemala area of MesoAmerica, and a second center of independent origin is in the Andean region of South America. In the Sierra Madre Occidental of Mexico, *Phaseolis vulgaris* grows in the central highlands from 10 to 25° N latitude, between 800 and 3000 meters elevation, where winter frosts are common (Kaplan, 1981). Characters separating domesticates from their wild ancestors include (1) extensive regional diversity of races with different seed colors, patterns, and shapes; and (2) relatively large seed size. In both modern farmers' markets in central Mexico and in archaeological sites dating more than 950 BP in the Tehuacan Valley of Mexico, four to six cultivars of vining beans are commonly found intermixed. The plants are self-pollinated (cleistogamous), which preserves genetic diversity once established. In the mixture of bean varieties, germination times tend to be staggered, which helps to ensure crop success in an uncertain environment, optimizing yield even at the expense of a single variety's potentially higher yield when sown alone. Today, in Mexico beans are typically grown with maize, planted before spring rains. Kaplan (1981) suggested that humans did not necessarily originate the components of varietal mixtures but assembled them from pre-existing variants into mixtures commonly used in planting.

The archaeological record contains no documentation of a long-term trend toward increasing seed size in the common bean. Instead, beans are consistently large throughout the archaeological record. In natural populations, large beans with thin seed coats are selected against because of seed predation by weevils. This can be countered if the beans are parched, for example heated to 60 °C in warm sand next to a hearth fire. Janzen (1969) speculated that human intervention freed the common bean from its co-evolutionary size constraint, and allowed for a rapid transition from small to large size. Because this morphologic transition is not observed directly in any one archaeological sequence, it must have come early, perhaps prior to 11 500 BP in South America, and certainly by 7800 BP in Mexico, during a cultural period of food gathering or early cultivation in which storage methods that eliminated predation by weevils permitted natural selection to operate on an existing polygenetic system.

EVOLUTIONARY ECOLOGY OF THE PLANT DOMESTICATION PROCESS

Winterhalder and Goland (1997) used foraging theory to understand the circumstances that led humans to select particular plants, and to look at the economic and population processes accompanying dependence on domesticated plants. Rather than ascribing changes in the utilization of food plants to changes in human population size, climate change, technological innovation, or efficiency of extracting energy from resources, they developed a perspective based on evolutionary ecology theory to look at a series of trade-offs between diet choice and subsistence risk.

The plant domestication model of Winterhalder and Goland (1997) is based upon foraging strategies that emphasize the consequences of localized decisions about resource selection. Their model thus shows how decisions about selection of resources could bring foragers into contact with plants that potentially could be domesticated. This model is complementary to the perspective of Rindos (1984), who asserted that plant domestication developed as a coevolutionary process between humans and plants, as humans at first were unintentional agents of seed dispersal and protection while harvesting plants, then underwent behavioral changes as they began to deliberately select for desired morphological characteristics of their food plants.

Winterhalder and Goland (1997) first assumed that potential plant resources may be present in the immediate environment, but at a population threshold too low to be initially included as an important part of the human diet. The relative value of ranking of the plant resource is established by its desirability as a food item. If the abundance of some more highly ranked species declines, the "transitional domesticate" species may enter the diet, and subsequently is subject to coevolutionary pressures (Rindos, 1984) that affect both its profitability (ease of harvesting or processing, as in increases in the size of seeds or thinning of seed coats) and its density (which may be increased by tilling, weeding, clearing a patch of soil, or fertilizing the soil). When such a plant first enters the diet it is of little importance to the subsistence economy, and makes up only a small proportion of foods used. Even as it coevolves to a position of low density and high rank, the selection of other food items in the diet may not change, and the density of foragers may not increase. Therefore, even with some human management, low-density resources may be incorporated into the diet without any appreciable change in the hunter–forager economy ("incidental domestication" of Rindos, 1984).

If, however, the transitional domesticated plant is initially very abundant, it will still be ignored until overall foraging efficiency declines enough (for example, because of climate change or cultural over-exploitation of this resource) to include it in the optimal diet. At that point, its economic impact will be high, and high yield will result in human population growth. With continued exploitation of high-ranking resources, depletion of resources may occur by over-exploitation, and consequently the diet breadth may eventually become more narrow as some resources become locally extinct and the transitional domesticate species assumes a major role in the diet. As this happens, the human population becomes more vulnerable to stochastic fluctuations in the populations of the newly domesticated species. In the terminology of Rindos (1984), such a sudden dependence on a new domesticate is the beginning of "specialized domestication."

The last hypothetical example of Winterhalder and Goland (1997) is one in which the profitability of the transitional domesticate plant increases through time, and attains both a high rank and high density. Such specialization leads to high risk as the human population becomes dependent on the newly domesticated plant.

The species that were to become the first native domesticates in eastern North America were low-density resources that remained part of a broad foraging spectrum (Kelly, 1995; Winterhalder and Goland, 1997). Archaic and Woodland people used these plants as supplements to a widely diversified diet while relying on nuts and animal resources for most of their caloric intake. Native crops may have been most important as a late winter–early spring supplement when other food resources were low (Smith, 1989). This is confirmed by paleofecal analysis from Salts Cave and Mammoth Cave dating between 2800 and 2350 BP (Gremillion and Sobolik, 1996). In the Woodland period, visitors to these cave sites ate the seeds of goosefoot, sunflower, marsh elder, and squash as well as hickory nuts not only during the growing season but also during winter and early spring. These native plant crops were integral parts of the human diet; frequent consumption of these foodstuffs "out of season" is evidence that they were storable resources, and this may have been an important determinant of their importance and selection as food.

MAIZE DOMESTICATION AND THE DEMISE OF FORAGING STRATEGIES

Morphological and molecular evidence pinpoints the New World origin and subsequent evolutionary history for races of domesticated maize (*Zea mays* subspecies *mays*) and closely related plants of teosinte (*Zea mays* subspecies *mexicana* and *parviglumis*) (Doebley, 1994; Goloubinoff *et al.*, 1994). From its evolutionary heartland in the Mexican highlands and in Central America (Wilkes, 1989; González, 1994), humans dispersed maize into coastal lowlands of northern South America well before 7800 BP, possibly reaching the southern Andes of Argentina as early as 10 200 BP (Pearsall, 1994). Introduction of maize into the American Southwest is recorded by direct dating of prehistoric remains of corn cobs and kernals, first known in southern New Mexico about 3400 BP, and in southern Arizona by 2900 BP (Adams, 1994). Migration routes may have been either from the Pacific coast of Mexico through northwestern Mexico and then into the southwestern United States, or northward from the highlands of central Mexico. Nearly a dozen major varieties of maize were developed in the American Southwest over the subsequent two millennia. By the late Archaic period, maize agriculture was established in southern Arizona, portions of New Mexico, and north as far as Utah.

For the Eastern Woodlands, a major change in foraging strategy occurred after the introduction of maize, which is dated to about 1950 BP in the American Bottom of southern Illinois (Fritz, 1999); 1700 BP in the Little Tennessee River Valley of eastern Tennessee (Chapman and Crites, 1987); and 950 BP in the Great Lakes region (Crawford *et al.*, 1997). Visibility of maize in eastern archaeological contexts increased dramatically about 1000 BP. Within two centuries field agriculture, centered on extensive cultivation of maize, emerged across much of the Eastern Woodlands, with the addition of squash and beans by 650 BP. Maize may have been an experimental or garden crop through the first 400 to 600 years after its introduction to eastern North America, but after 1000 BP it was planted extensively and dominated the food production system.

Maize is a prime example of an introduced food species that became both highly ranked and was a potentially dense resource (Winterhalder and Goland, 1997). As human populations increased in density in the Mississippian period (from about 1000 to 500 BP), maize dominated the diet. As demonstrated by stable carbon isotope studies (Lynott *et al.*, 1986; Schoeninger and Schurr, 1994; Greenlee, 1998), the proportion of maize in the diet increased rapidly after 1150 BP, marking a quick transition of maize from a low-ranked to a high-ranked food resource. This transition may have occurred because of technological innovations in food production, or because of the development of new varieties of maize. Maize entered the diet of eastern North Americans as a low-ranked food item. Other kinds of starchy and oily seeds continued to be used through the first half-millennium during which maize was available.

Available supplies of other, more highly ranked native plant food resources may have been over-exploited, however, as human populations increased. Only after the Mississippian dietary shift to maize did charred hickory and walnut wood show up in large quantities within preserved hearths. This transition in cultural preferences for plant selection may track building population pressures. Rindos (1989, p. 32) concluded that "The growing need for land on which to grow maize to support a burgeoning population probably brought about the destruction of what had previously been an important dietary resource – the mast crop harvested from nut trees." As stated by Winterhalder and Goland (1997, p. 149),

> Risk will be especially high under conditions . . . in which a dominant high-density/low-ranked resource is accompanied by broad diet breadth, population increase, and depletion of highly ranked resources. Early farming people such as Mississippian populations may have faced greater risk at the same time that they had fewer strategies for buffering shortfalls, on account of population increases,

reliance on fewer crops, and a diminished array of wild fallback resources because of environmental deterioration resulting from overexploitation and land clearing.

CONCLUSIONS: THE TRANSITION FROM FORAGERS TO FARMERS

In North America, the transition from foragers to farmers occurred over a long time span that was characterized by a number of facilitative interactions among plants, their pollinators, their human seed dispersers, and their harvesters. During the Archaic cultural period, throughout the Eastern Woodlands, Native Americans continued to practice lifeways based upon a mixed economy in which plant foods were an important supplement to game and shellfish. Incipient domestication of native plant species occurred primarily as a result of human disturbance of riparian corridors, where their semi-permanent residential sites opened up light gaps suitable for colonization and spread of weedy ruderals. Initial domestication of both exotic and native food plants varied in intensity and sequence across the Eastern Woodlands.

In the process of becoming farmers, Archaic Native Americans altered the ecological setting, primarily through their mutualistic relationships with individual plant domesticates, and by encouraging the spread of riparian weeds that thrived in dooryard gardens of Archaic village sites along major river systems of the Eastern Woodlands. With domestication of plants such as squash, gourd, sunflower, goosefoot, and marsh elder, as well as maize and beans, biological diversity was enhanced at the within-species level. This selective effect was temporary for goosefoot and marsh elder that reverted to their wild types after these species were no longer propagated. For maize, cucurbits, and beans, however, prehistoric cultivation has left a lasting legacy of genetic diversity. In the case of squash, its range of modern phenotypic plasticity exemplifies the evolutionary–ecological role of prehistoric Native Americans during the mid- and late Holocene.

5

Population-level interactions

HUNTER–FORAGER ADAPTATIONS TO THE PLEISTOCENE/HOLOCENE TRANSITION

Changes in human subsistence patterns occurred in response to new opportunities presented by the change from Pleistocene to Holocene climate, vegetation, and fauna. By 11 500 BP, a foraging type of ecosystem type developed in which subsistence systems were becoming diversified, with a switch to white-tailed deer (*Oedocoilus virginianus*), moose (*Alces alces*), and woodland caribou (*Rangifer tarandus*) as primary prey animals in northern and eastern North America, and more intensive exploitation of local resources, including small game, nuts, fish, and shellfish that were previously not a major part of the prehistoric human diet (Stoltman and Baerreis, 1983). Because human populations had spread throughout North America, no new territories were left to colonize, and instead people developed localized technologies reflecting increasingly intensive use of local resources, including milling and nutting stones, steatite stone and ceramic containers, and fishing gear such as harpoon heads, net weights, and hooks. Social boundaries developed between group territories, and permanent communities were established (Stoltman and Baerreis, 1983).

Detecting the ecological interactions of prehistoric hunter–gatherers requires understanding the landscape context in which they interacted at the close of the last Ice Age (Yesner, 1996). One of the more important considerations is the size of the grain of habitats relative to the extent of the landscape occupied (Guthrie, 1984; Turner *et al.*, 2001).

Spatially explicit models of settlement patterns

Spatially explicit models of settlement patterns relative to resource utilization are relevant to interpreting Paleoindian and Archaic period hunter–gatherer adaptations. Three alternative models for broad-scale (biome-level) settlement patterns of prehistoric hunter–gatherers are based upon location strategies, mobility patterns, and decision-making factors (Butzer, 1982). The simplest model assumes that the environment was uniform and that human settlements were dispersed randomly across biomes. A more complex model recognizes that spatial patterning in plant productivity and animal biomass was important, leading to clustering of human occupations within biomes. Discontinuities in resource availability at biome boundaries would have affected settlement strategies, which ultimately reflect cultural responses to a complex and variable ecological gradient.

Medium-scale settlement patterns for hunter–gatherers can be viewed in the context of a model of human adaptive responses to either dispersed or clustered resources. Where water is abundant and plant and animal foods are predictable and well dispersed, hunter–gatherers would tend to occupy discrete, equidimensional territories. Where food resources, water, and topographic features are diffuse, human territories would be diffuse. Alignment of a group of settlement networks would tend to occur along the seashore, a mountain valley, rivers, or a line of springs as controlled by topography. A radial pattern of settlement might be observed around a circular or elliptical resource base, such as a lake, marsh, or spring-fed oasis. Overall, the population of a human group is dependent on resource dispersal, and on aggregation and distribution of all other human groups in a fixed area (Butzer, 1982).

Optimal foraging theory

Optimal foraging theory, initially developed for ecological analysis of predator–prey systems, forms a basis for understanding human hunter–forager resource strategies in environmentally variable space (MacArthur and Pianka, 1966; Winterhalder, 1981). In a uniform environment where resources are abundant, human settlements should show a regular pattern of dispersion of the smallest viable social units. Clustered resources, however, favor aggregation of social units at a central place. Hunter–gatherers living in a habitat with fine-grained patch types would tend to make generalized use of available patch types. Those living in a large-grained habitat would tend to become specialized in their use of patch types. Human diet breadth would be influenced by the abundance of resource items ranked high in terms of efficiency of hunting or gathering activities. The number of patch types exploited in the course of food procurement will affect the cost of time for searching, collecting, and harvesting yields. When resource density increases, the number of patch types included in foraging activities can be decreased (Winterhalder, 1981).

Landscape complementation and supplementation

Optimal foraging theory can be extended using landscape–ecological principles. Ecological processes operating at the landscape scale – that is, intermediate between an organism's normal home range and its regional distribution – are incorporated within two integral features of landscape structure: (1) the composition of habitat types; and (2) the spatial arrangement of habitats (Turner, 1989). Two kinds of basic ecological processes that can influence population dynamics or community structure and that are relevant to consideration of the activities of Archaic hunter–foragers are landscape complementation and landscape supplementation, both of which occur when individuals move between patches in the landscape (Dunning *et al.*, 1992). In the case of landscape complementation, a species requiring nonsubstitutable resources found in two habitats can support larger populations if the two kinds of resource patches are located in close proximity to each other rather than far apart. For humans, examples of nonsubstitutable resources might be foraging patches and base-camp sites (Binford, 1980). The benefits of close proximity of resource patches are two-fold – not only is it more efficient to glean resources, but also the forager can minimize the length of time exposed to predators while traveling between patches.

Organisms that utilize different kinds of resource patches can also benefit from substitutable resources. Landscape supplementation is a process by which organisms supplement their resource intake either by using resources in nearby patches of the same habitat or by substituting another resource in nearby patches of a different type. Thus, a population can persist in a patch that is too small to contain all the resources the population needs (Dunning *et al.*, 1992). In the Eastern Woodlands, Archaic humans foraging for acorns and hickory nuts may have relied on several stands of trees for sustenance of a small band, particularly given the localized dispersal of nuts and the variability in nut production from grove to grove resulting from multiple-year cycles in mast production. During poor oak mast years, alternative sources of nutmeats may have been sought in walnuts, chestnuts, or hazelnuts (Talalay *et al.*, 1984; Petrúso and Wickens, 1984). Another example of landscape supplementation is in the use of rock outcrops located long distances away from base camps for procurement of high-quality chert for projectile point-making (Tankersley, 1998).

Logistically organized collectors

Hunter–gatherer subsistence strategies may also be described in terms of different kinds of social organization in which the activities of foragers are very different from those of "collectors" (Binford, 1980). The foraging system is based upon establishing a residential base within a favorable location. Foragers typically do not store food, but gather food items daily and return to their base camps at night. The seasonal round of activities involves a series of residential moves among resource patches. In large or homogeneous resource patches, the number of residential moves may increase, but the distance between moves is minimized. If, however, resources are scarce and dispersed, the size of the mobile group may be reduced and the family units scattered over a large area.

In the collector strategy, rather than taking the people to the resources, the resources are brought to the resident population by a series of "logistically organized collectors" – special task groups organized to procure food by establishing field camps in addition to the primary base camp (Figure 3.3.). Each kind of field camp is a temporary operating center where food is procured and raw materials may be processed, for example kill sites for game, and fish weirs. In addition, stations may be set up for the purpose of gathering information, for example about the movements of game or other people. Caches are a third kind of site used by collectors, for temporary storage of excess food resources. Forager–gatherers use strategies to organize their activities. These logistical forays are directed toward specific resources, such as fish or salt or high-quality chert, that are critical but located far from other critical resources (Binford, 1980).

The archaeological record

Models of hunter–forager strategies can help elucidate the processes behind the patterns of tangible artifact remains that constitute the archaeological record (Binford, 1980). For the period during and after the transition from Paleoindian to Archaic cultural periods (11 500 BP; Figure 3.5), the archaeological record is consistent with a diffuse subsistence strategy rather than a highly specialized "focal" strategy (Meltzer and Smith, 1986). During mid- to late-Holocene times, however, a subtle shift in human diet took place, with the number of plant and animal species included in the diet increasing, including riverine aquatic species. Rather than reflecting the development of a more efficient seasonal round of foraging activities, however, this dietary shift may have resulted from adjustments to a decreasing resource base as human populations increased and used smaller areas of habitat more intensively (Meltzer and Smith, 1986).

EVOLUTION OF INTERACTION BEHAVIOR

Late-Pleistocene bands of Paleoindians had to interact with other bands in order to survive (Anderson, 1995a, b; 1996). Evolution of new modes of interaction behavior established the regional social templates that structured subsequent Holocene cultural evolution across North America (Kowalewski, 1995). Exploration of an unknown and potentially hostile New World posed critical problems to newly founded groups of Americans. These groups had incomplete knowledge about regional resources and yet were motivated by vulnerable biological imperatives such as finding mates, and surviving crashes in subsistence resources such as megafauna (Sassaman and Nassaney, 1995).

Staging areas and communication hubs

With expanding population growth and increased interband competition for limited resources, patterns developed in dispersal of human populations that led to aggregation in staging areas and to occupation of more-or-less bounded territories. Paradoxically, the need to find suitable mates of appropriate age and kinship distance (to minimize inbreeding) required social interactions that brought people together. However, with population growth other social interactions were required to help space out the bands, effectively dispersing people across the landscape to new strategic locations. Small mobile bands of people maintained social ties for exchange of goods and information in order to monitor resource availability, identify unoccupied lands, and avoid redundant land use associated with human saturation of available territories.

Mobility strategies of Paleoindian groups changed along with changes in their intergroup interactions (Anderson, 1995b). Between 15 000 and 11 500 BP, from early to late Paleoindian times, a four-fold increase in archaeological site density occurred. The increasing number of sites during that interval is evidence of initial human exploration of previously unsettled areas followed by waves of colonization. Late-Paleoindian people occupied permanent settlements in "staging areas" centrally placed within resource-rich environments along major riverways, coastal shorelines, mountain ranges, and biome boundaries. Broad regional cultural differences reflected divergent societal adaptations to the Eastern and

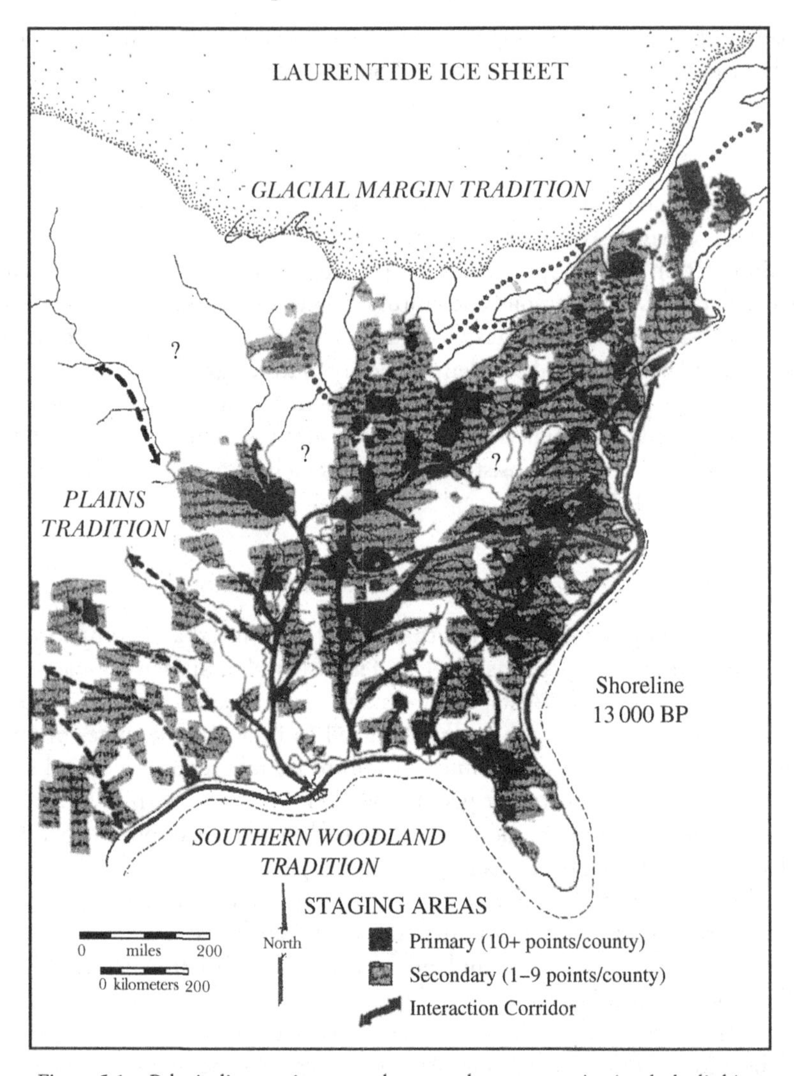

Figure 5.1 Paleoindian staging areas that served as communication hubs linking interaction corridors and fostered the cultural emergence of regional adaptive lifeways from 13 000 to 12 500 BP. The light and dark shaded areas reflect the intensity of Paleoindian occupation in terms of the number of archaeological sites containing diagnostic stone projectiles (modified from Anderson, 1995b).

Western Woodlands, to Great Plains prairies, and to northern tundra/taiga biomes (Figure 5.1). Other settlement patterns of macro-band clusters radiated outward to the hinterlands from different staging areas, linking "communication hubs" along interaction corridors and emerging centers of regional cultural traditions. Landscape corridors funneled group movements and facilitated scheduled rendezvous at seasonal times of abundant food resources for fish, waterfowl, and white-tailed deer (Anderson, 1995b).

Macro-band networks

Over the Pleistocene/Holocene transition, the free-wandering Early Paleoindians may have established predictably extensive mobility patterns with a seasonal round of base camps in a broad geographic, although vaguely defined, territory (Anderson, 1995b, 1996). Late Paleoindians and early Archaic people developed networks of interacting macro-bands that were seasonally tethered to primary camps and centrally located within fixed territories. Rapid population growth coincided with geographic expansion, with new bands forming by fissioning into new peripheral territories. This fissioning in turn reduced the effective size of spatially bounded ranges, socially isolating more distant outlying colonies. Reductions in territorial size would be tied not only to the cultural perception of the size of range and landscape heterogeneity necessary to maintain group size, but also to the need to ensure adequate spacing of individual bands. The settlement network of related bands would promote a predictable frequency of interaction through the communication hub of the central staging area, maintaining a subregional population for finding mates, yet avoiding intergroup interference and depletion of game, plant food, and high-quality lithic resources (Anderson, 1995b, 1996). These theories can be tested using the spatial and temporal distribution of stylistically distinctive tool assemblages recovered from archaeological sites and placed within a paleoenvironmental context based on geological and paleoecological evidence.

THE CENTRAL MISSISSIPPI VALLEY

The alluvial corridor of the Central Mississippi Valley provided a primary invasion route for Paleoindian entry into southeastern North America (Anderson and Gillam, 2000). Bounded to the west by the Ozark Plateaus and to the east by the Blufflands, the Mississippi River Valley is further separated into Western and Eastern Lowlands by the north–south oriented, 75-meter-high Crowleys Ridge. In the Western Lowlands, relict braided-stream terraces consisting of river-deposited sand and gravel are mantled with thin loess and are today occupied by xeric oak–hickory forest. Sand dunes are preserved on knolls representing former point bars in a Pleistocene braided stream. Slack-water streams drain southeastward from the Ozark Border, with ash–willow communities occupying relict braided-stream channels (Delcourt *et al.*, 1999).

Late-glacial environments (16 800 to 11 500 BP)

During the late-Pleistocene/Holocene transition, several key changes occurred in landscape dynamics within the Central Mississippi Valley (Delcourt and Delcourt, 1996; Delcourt *et al.*, 1999). Between 16 800 and 14 000 BP, there was a progressive loss of braided-stream habitat as meltwater input to the Mississippi River from the retreating Laurentide Ice Sheet diminished east of Crowleys Ridge (Figure 5.2a). Incision of

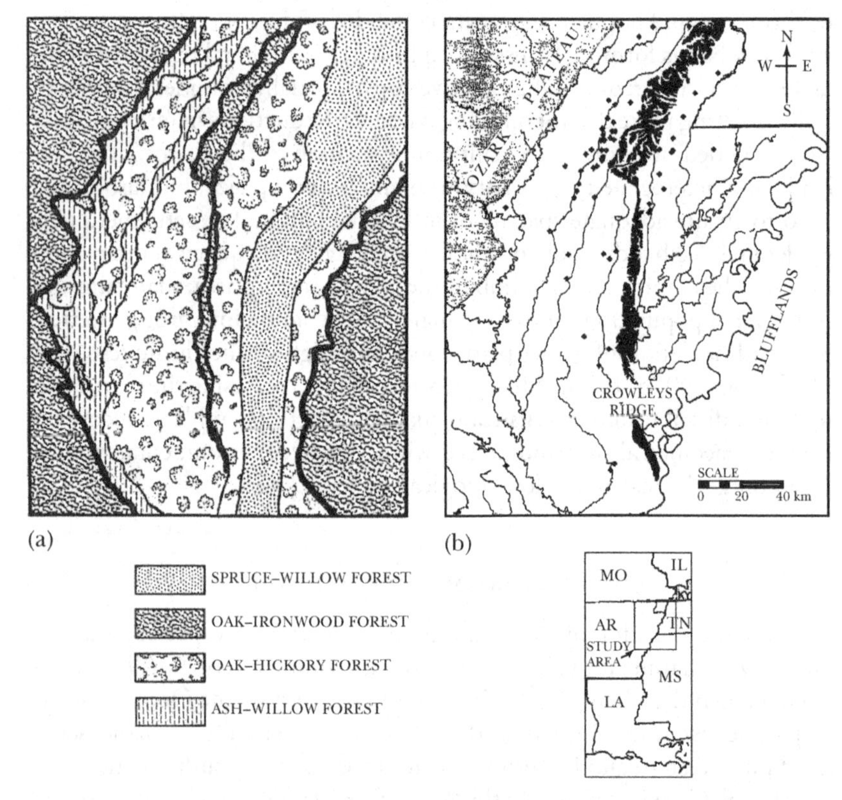

Figure 5.2 Late-Pleistocene vegetation patterns and distribution of Early Paleoindian sites (shown as dots) dating from 14 000 BP in the Central Mississippi Alluvial Valley (modified from Delcourt *et al.*, 1999 and Gillam, 1999).

braided-stream deposits reduced the area of point-bar surfaces that previously were the sources for sand and silt reworked by wind into dunes and loess, respectively. By 11 500 BP, the vegetation in the Eastern Lowlands changed from spruce–willow woodland to cypress–tupelo gum forest in permanent backswamps, to sweetgum–elm forest on seasonal floodplains, and to willow thickets and canebrakes along the levee crest of the meandering Mississippi River channel (Figure 5.3a).

In the Western Lowlands, glacial-age environments included active sand dunes on point-bar knolls deposited by glacial meltwater and colonized by patchy spruce–willow stands. By 14 000 BP, however, boreal-like vegetation had been replaced by ash–willow forest in relict braided-stream channels

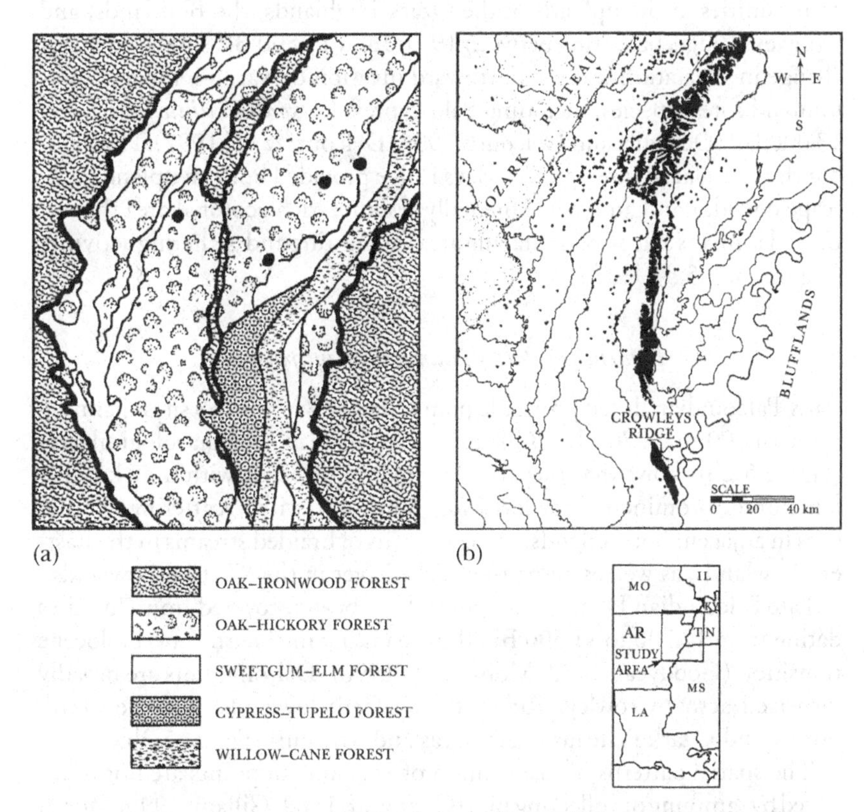

Figure 5.3 Vegetation patterns and distribution of Late-Paleoindian Dalton sites dating from 11 500 BP in the Central Mississippi Alluvial Valley (modified from Delcourt *et al.*, 1999 and Gillam, 1999). Fossil pollen sites are shown as dots in Figure 5.3a and Dalton archaeological sites are shown as dots in Figure 5.3b.

(Figure 5.2a). By 11 500 BP, ash and willow were replaced by modern plant communities including sweetgum–elm forest on seasonally flooded sites (Figure 5.3a).

Three key elements in the landscape dynamics accompanied the last major hydrologic changeover of the Mississippi River from braided regime to meandering river at about 12 500 BP. The first of these was the loss of braided-stream habitat and extinction of Critchfield spruce (*Picea critchfieldii*) at about 13 000 BP (Jackson and Weng, 1999), resulting in the loss of an early-successional patchwork of boreal-like vegetation that previously would have attracted mammoths and other now-extinct megafauna such as paleollama (*Palaeolama mirifica*; Morse and Graham, 1991). The second landscape change resulted from the northward invasion of temperate forest communities in the uplands of the Ozark Highlands, the Blufflands, and Crowleys Ridge, occurring between 19 100 and 11 500 BP. The third major change in vegetation was the northward invasion of warm–temperate bottomland forest species, including bald cypress (*Taxodium distichum*), after 13 000 BP (Delcourt and Delcourt, 1996; Delcourt *et al.*, 1997).Throughout the late-glacial interval, Crowleys Ridge provided both an upland landscape corridor and an environmentally resource-rich ecotonal area that attracted humans as a primary staging area for hunting and gathering activities (Figures 5.2, 5.3).

Evidence of Paleoindian settlement patterns

Early Paleoindian fluted projectile points are known from 68 sites spanning from 14 000 to 12 700 BP (Gilliam, 1999). A map of archaeological sites (Figure 5.2b) shows that they are centered on or occur within 30 km distance of the prominent Crowleys Ridge landform, with a scattering of other sites in adjacent bottomlands, at the margins of braided streams in the Eastern Lowlands, as well as along the Cache River in the Western Lowlands.

Late Paleoindian Dalton-style points have been recovered from 767 sites dating from 12 700 to 11 500 BP, thus spanning the Pleistocene/Holocene transition (Goodyear, 1982; Morse *et al.*, 1996). Dalton points are broadly dispersed across Crowleys Ridge, the adjacent lowlands, and the Ozark border and Ozark Plateaus of Arkansas and Missouri (Figure 5.3b).

The spatial patterns of distribution of Paleoindian points are not influenced by sampling or collecting bias (Goodyear, 1982; Gillam, 1999; Morse, 1997). Rather, the sites are distributed according to proximity to major resources of raw lithic material on Crowleys Ridge, or to plant and animal food resources along floodplains of major streams. Chert nodules constituting the raw lithic resources that were potentially important for manufacture of stone tools are patchily distributed in the Central Mississippi Valley, occurring as chert-rich gravels that form Crowleys Ridge, channel-lag in gravel beds of streams flowing from Crowleys Ridge, and as chert nodules within limestone and dolomite exposed in the Ozark Plateaus. Overall, there is a significant association of Paleoindian sites with lithic sources on Crowleys Ridge, with Early Paleoindian sites concentrated primarily within 15 km, and with Dalton sites occurring significantly within 25 km. A second significant association is between Paleoindian sites and stream corridors, with many fluted-point sites located within 5 km of streams on outwash terraces capped by sand dune knolls.

Dalton sites occur in significant numbers only within 1 km adjacent to modern stream courses, but up to 4 km from streams in the terrace-edge zone. Overall, prehistoric settlement preferences corresponded with those of modern population centers, with a bimodal distribution reflecting (1) the Early and Late Paleoindian pattern of chert exploitation centered on the Crowleys Ridge landform; and (2) long-term settlement focused on the major river systems in the immediate vicinity of Crowleys Ridge, where aquatic habitat for fish and waterfowl included seasonally inundated bottomlands and remnant braided-stream terrace surfaces with a patchwork of open, active sand dunes and xeric oak–hickory forests.

Gillam (1999) envisioned long-term settlement of base camps along the bottomland corridor in Dalton times, with spring and summer harvest of aquatic and terrestrial foods. Short-term fall and winter base camps were located on sand dunes of high braided-stream terraces, where nut mast and white-tailed deer would have been abundant in patchy oak–hickory forests. Short-term forays would have been made to replenish tool kits at stone quarries on Crowleys Ridge.

Regional cultural adaptation and population increases tracked changes in landscape configuration and potential food production from deciduous upland forests and floodplain habitats. Increased knowledge of local resources by Dalton people resulted in geographic saturation of the Central Mississippi Valley and bordering uplands, with the creation of more bands but with decreased overall range of seasonal group movement.

Test of spatially explicit models of Dalton settlement

Schiffer (1975a,b) proposed that band-level territories would have crosscut drainages, spanning significant ecotones in order for Paleoindians to exploit contrasting environmental zones. He envisioned that territories thus established would be equidistant, allowing hunter–gatherers to saturate all portions of the landscape. In this context, base camps would be frequently reoccupied, and more ephemeral hunting forays would be "archaeologically invisible." Seasonal camps would have been moved continually across the whole watershed and onto one or more upland areas bordering the Western Lowlands (Schiffer, 1975a,b).

An alternative model is one in which bands of Dalton people were centered within particular portions of single drainage areas (Morse, 1971). This model of settlement patterns projects that at least three bands of Early Paleoindians may have preferentially settled in the upper Cache River and Crowleys Ridge, with a total of at least seven territories distributed across the Western Lowlands. Each band-level territory would have included centrally located, long-term base camps, related cemeteries, and seasonal short-term logistical forays for hunting and gathering camps and trips to stone quarries on Crowleys Ridge to replenish tool kits.

Support for Morse's (1971) settlement model is derived from technological innovations made by Dalton people. The Dalton technological development of chert adzes, cutting tools hafted onto wooden shafts, represented the transition to a lifeway that included deliberate shaping of wood items (Morse, 1997), including the making of dugout canoes from bald cypress (*Taxodium distichum*), a tree that was established in the Central Mississippi Valley by about 13 000 BP (Delcourt *et al.*, 1997). Travel by canoe would have meant greater mobility and preferential utilization of riparian corridors.

Sources of chert used for Dalton points also support the Morse (1971) settlement model. A large cache of Dalton points, recovered from the oldest known cemetery in the New World, dates to 12 500 BP. The Sloan Cemetery is located on a sand dune bordering the Cache River. Along with human burials, 439 chert artifacts were found that came primarily from local sources on Crowleys Ridge but that also included rarer cherts from outcrops 75 km to the southwest on the Ozark Plateau and 300 km to the northeast near St. Louis, Missouri. The artifacts from the Sloan Cemetery demonstrate that the home range of Dalton people would have been within 50 km of their central base camps in the Western Lowlands, but that they took occasional, short-term forays as far as several hundred kilometers to quarries elsewhere along the Mississippi Valley (Morse, 1997).

Ecological implications of Paleoindian population estimates

Based on the available archaeological evidence, at the Pleistocene/Holocene transition around 11 500 BP, Paleoindians in the Central Mississippi Valley were too few in number, too limited in stone tool technology, and too dispersed in their seasonal round of hunting and foraging to have played a substantive ecological role as a keystone species.

Long-term Dalton settlement was primarily focused on major rivers in the alluvial lowlands bordering Crowleys Ridge (Morse, 1997; Gillam, 1999). Dalton bands were ultimately bound together along the watershed they occupied, all branching out in a network tributary to the Mississippi River. Each band or local group would have maintained a "critical population size" with four active mature males (enough to conduct a drive ambush for white-tailed deer) and two to three active, reproductively mature females who gathered wild plants and ensured longevity of the group.

Assuming that band size included 25 to 50 individuals, and with clusters of archaeological sites indicating that three Early Paleoindian bands lived in northeastern Arkansas, their total population density would have been only 75 to 150 people within a 30 000 km^2 territory. After another thousand years and formation of as many as seven Dalton bands, the total human population may have doubled within the Central Mississippi Valley, growing to a macro-band size of 175 to 350 individuals. Dispersing further into the upland hinterlands, other Dalton bands may have formed new staging areas along the tributary streams draining the Ozark Plateau (Wyckoff and Bartlett, 1995).

THE PRAIRIE/FOREST BORDER

The fossil pollen record from Ferndale Bog in eastern Oklahoma (Albert and Wyckoff, 1981; Bryant and Holloway, 1985) shows that late-Pleistocene boreal-like woodland was replaced by prairie vegetation by 13 000 BP. A major ecotone between forest and grasslands has thus been in place through central Oklahoma since Clovis times, resulting in a biologically rich transition zone between biomes containing a mixture of both eastern and western plant and animal species.

A fundamental cultural border that developed during the Pleistocene/Holocene transition (Figure 5.1) coincided with the development of the ecotone between temperate eastern deciduous forest and temperate grassland in the southern Great Plains region of Oklahoma (Wyckoff and Bartlett, 1995). Radiocarbon dates from Folsom sites in the Southern High Plains range from 12 900 to 11 800 BP. Folsom points are concentrated in the western half of Oklahoma. Folsom people were primarily bison hunters. Western Oklahoma was on the northeastern periphery of a "Grand Circle" of seasonal activities. Folsom people stocked their chipped-stone tool kits with points made from Cretaceous cherts from the Edwards Plateau in south-central Texas, then moved north to the High Plains to hunt bison (*Bison bison*). They replenished their tool kits with agatized dolomite from quarries in northwestern Oklahoma, then traveled east through the Osage Plains and south to Texas. The Oklahoma area was thus on the eastern periphery of the Folsom hunting range.

Archaeological evidence from Rodgers Shelter, Missouri, indicates that Dalton occupation dates as early as 12 500 BP in the western Ozarks (McMillan, 1976). Folsom and Dalton cultures were therefore contemporaneous between 12 500 and 11 800 BP (Wyckoff and Bartlett, 1995). Unlike in the Folsom culture, however, the Dalton tool kit was specialized for hunting white-tailed deer and small mammals that lived in forest and forest-edge habitats. The distribution of Dalton points and adzes made of stone from the western Ozark and Ouachita highlands is concentrated along westward-flowing streams in the Osage Savannah region of Oklahoma.

It remains unknown whether Folsom people interacted with Dalton people where their territories overlapped in the Osage Plains of central and western Oklahoma. Only a few Folsom points have been recovered from eastern Oklahoma, and few were made of materials quarried in the interior highlands of the Ozarks or the Ouachitas. Population levels probably remained low throughout the 35-generation (700-year) overlap in the two cultures.

The prairie/forest border functioned as a late-glacial/early-Holocene ecological buffer zone, with access by many people from both the interior highlands and the plains. Folsom and Dalton people, however, were adapted to exploit different landscape mosaics along this major biome boundary (Butzer, 1982). The Osage Plains of Oklahoma were on the periphery of the Folsom elliptical bison-hunting range. For Dalton hunter–gatherers, central Oklahoma was a homeland where they depended upon the variety of resources, moving east–west along riparian corridors and hunting and gathering in both gallery forests and open glades habitats. Despite some overlap in diagnostic artifacts, there is little evidence for direct interaction between the two groups, described by Wyckoff and Bartlett (1995) as "Paleoindian macroregions in spatial and temporal transition."

ROCKSHELTERS OF THE HINTERLANDS

Dalton people were the first to exploit rockshelter environments in the region from Missouri to the southern Appalachian Mountains (Walthall, 1998). For example, data from Rodgers Shelter, a shallow cave at the foot of a bluff overlooking the Pomme de Terre River in western Missouri, show that Dalton-period occupations were repeated seasonal visits made by small families who camped in the shelter from fall to spring. In the rockshelter, they worked animal hides and manufactured tools from local river cobbles.

In many rockshelter sites, faunal remains associated with Dalton artifacts typically include white-tailed deer and small mammals. Late Paleoindian hunter–gatherers such as Dalton people were highly mobile foragers who followed an annual round related to the distribution of resources across the landscape and through the seasons (Walthall, 1998). Dalton culture was characterized by high residential mobility, short-term fall gatherings of populations, and disperson of groups during the winter. People moved from summer camps in floodplains to mast-rich, forested uplands, including rockshelter sites, in the late fall (C. Chapman, 1975).

Occupation of rockshelters by Dalton people was not necessarily a direct result of increases in human population size that forced a migration to marginal habitats. Rather, "the widespread occupation of rockshelters by early-Holocene Dalton groups signals a fundamental reorganization of hunter–gatherer mobility strategies and establishes the basic annual cycle that is common to human adaptation in the region throughout the remainder of the Holocene, from Dalton to the ethnographic present" (Walthall, 1998, p. 231).

EARLY ARCHAIC ADAPTATIONS ON THE SOUTHERN ATLANTIC SEABOARD

The region of the southern Atlantic Seaboard includes the southeastern Blue Ridge Escarpment, the Piedmont, the Fall Line ecotone, and the Inner and Outer Coastal Plain of southeastern North America (Figure 5.4). After 11 500 BP, Early Archaic people formed incipient macro-bands offset by 250 to 400 km between staging areas. In South Carolina, these clusters of allied bands were concentrated near the Fall Line along the Congaree River and near the Allendale and Brier Creek chert quarries along the Lower Savannah River (Anderson, 1996).

The Fall Line represented a significant ecotone that was easily observable as a break in terrain from the hilly Piedmont with its deeply weathered, red soil to the gently rolling coastal plain underlain by white sand. As documented by paleoecological evidence from White Pond, South Carolina, forests 11 500 to 8900 BP would have been dominated by mesic hardwoods including American beech (*Fagus grandifolia*), oaks, and hickories on the Piedmont, with drier oak–hickory–southern pine forest predominating over much of the Coastal Plain (Watts, 1980a).

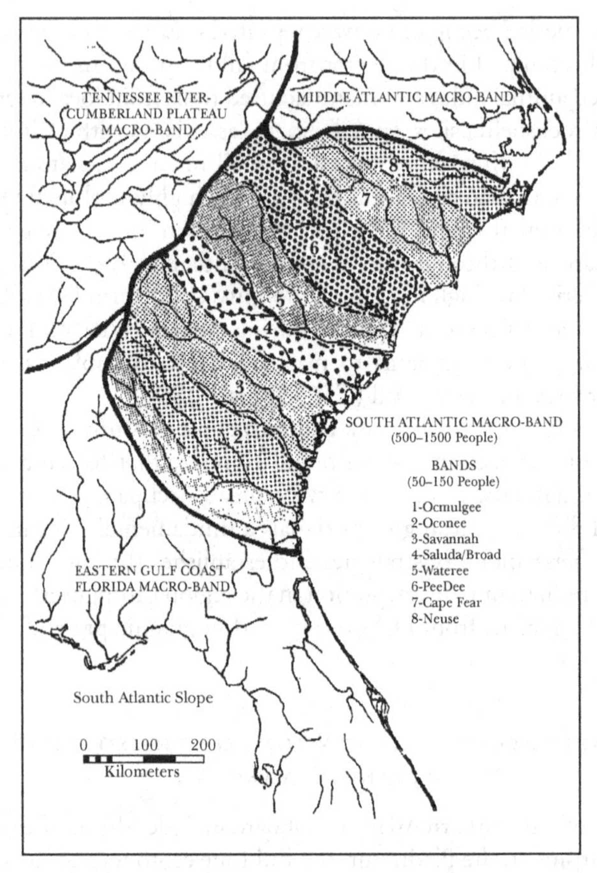

Figure 5.4 Map depicting the territory of the Savannah River Early Archaic macro-band (Number 3), one of eight macro-bands in the Southern Atlantic Seaboard region, each of which moved on an annual round of foraging along riverways between the Piedmont headwaters and coastal estuaries of Georgia and the Carolinas (modified from Anderson and Hanson, 1988).

Biocultural constraints

Early Archaic adaptations, between 11 500 and 8900 BP, were controlled by four factors, or "biocultural constraints" (Anderson and Hanson, 1988): (1) environmental structure and predictability of the seasonal and geographic pattern of available food resources; (2) cultural interaction for regulating and maintaining an interband network for finding suitable mates; (3) information exchange to influence social interaction and to assess resource quality and availability; and (4) demographic structure, including both population size and spacing.

Two levels of settlement organization have been inferred from the archaeological evidence (Anderson and Hanson, 1988). The regional macro-band social system was comprised of an aggregate of eight to ten bands, with a minimum equilibrium size for the mating network of 500 to 1500 people. Such a macro-band network would have encompassed much of the southern Atlantic Seaboard, from the Ocmulgee to the Neuse river basins (Figure 5.4). Within this broader social structure existed local bands, with 50 to 150 people utilizing a two-tier radius for foraging plants and animals within a residentially mobile forager–collector strategy (Figure 5.4). Local bands would have moved seasonally within one drainage basin, such as the Savannah River watershed, establishing logistically provisioned base camps in late fall and winter, with rendezvous at the Fall Line along the river and near the Allendale chert quarries. Short-term foraging camps would have been established throughout the rest of the year, in spring moving toward the coast and in late spring, summer, and fall relocating toward the Inner Coastal Plain and Piedmont, moving across drainage divides to join up with multi-band or macro-band gatherings (Anderson and Hanson, 1988).

Seasonal predictability of food resources

In winter, food resources would have been spatially patchy and variable on the Inner Coastal Plain (Anderson, 1966). Base camps would have been located along rivers, with field stations or logistical camps located in the Fall Line Sandhills or along stream terraces, where white-tailed deer would aggregate, or "yard up," during the winter (Sassaman, 1996). In the spring, warm weather would result in leaf-bud break of dormant plants and in dispersal of white-tailed deer populations, occurring first near the Atlantic coast and later on the higher elevations of the Piedmont. Archaic people may have had a generalist springtime foraging strategy as humans dispersed into small, regularly spaced camps that were frequently relocated, following the progression of spring weather from southeast to northwest across the Fall Line. The summer foraging strategy may have been constrained by the location of outcrops for lithic materials, located in the Piedmont (Anderson, 1996; Sassaman, 1996).

The archaeological evidence for Early Archaic adaptations

Evidence for occupation patterns in the Early Archaic period is primarily found in the lithic composition of distinctive tools such as hafted biface points, recovered from archaeological sites throughout the southern Atlantic Seaboard (Sassaman, 1996). Along the Savannah River, there was a cultural preference for high-quality Allenale chert, which was transported up to 150 to 200 km from its quarry source. In the array of archaeological sites situated along the Savannah River Valley, gradual changes occurred between adjacent sites in the types of raw material used for projectile points. The absence of spatial discontinuities in raw materials used indicates that Archaic people were not tethered to mutually exclusive portions of the watershed. The relatively low density of archaeological sites is interpreted to reflect limited, although increasing, population densities from 11 500 to 10 700 BP. The long-distance transport of coastal-plain cherts into the Piedmont is considered artifact evidence that bands of humans moved at least 150 km along their seasonal round of foraging. The Fall Line transect marking the major landform break between hilly Piedmont and rolling coastal plain served as a corridor for human movement across drainage basins, presumably for social gatherings and annual rendezvous. The significant transport of exotic lithic materials such as metavolcanic rock and the prized Allendale cherts is considered evidence for substantive human population movement crossing the grain of the landscape along this prominent Fall Line ecotone.

Early Archaic people utilized riparian corridors that focused their seasonal movements. Base camps were established along the major river floodplains and terraces, and resources were extracted within a linear foraging zone. Small field stations were widely dispersed across the upland interfluves, with long-distance hunting forays expanding into the peripheral logistic zone (Sassaman, 1996).

After 8900 BP, Archaic populations declined within the Atlantic Coastal Plain as they shifted to more permanent settlements within the Piedmont (Anderson, 1996). This spatial shift reflected the tracking of a change in the quality of forest composition. On the southern Atlantic and Gulf Coastal Plains, the rise to dominance of southern pine forest displaced oak–hickory forest in the mid-Holocene (Watts, 1980b; Delcourt, 1980a). Archaic people responded by moving from the Coastal Plain, attracted to Piedmont forests that provided a continuing source of mast nuts (Anderson, 1996).

MAST EXPLOITATION AND TREE POPULATIONS

In the Archaic cultural period, nut gathering became an important form of foraging prior to and during the transition to farming. In the Eastern Woodlands, exploitation of mast followed climate warming and the development of widespread deciduous forest during the early and mid-Holocene. With a shift toward warm, dry weather patterns, between 11 500 and 6800 BP oak (*Quercus*), hickory (*Carya*), and walnut (*Juglans*) trees expanded in geographic range and dominated regional forests across the eastern United States, from the riparian corridors of major rivers to the upper slopes of tributary headwaters (Delcourt and Delcourt, 2000). Pollen evidence substantiates increased pollen productivity for hickory and oak from approximately 9500 until 4500 BP in central Kentucky, Middle Tennessee, and southeastern Missouri (Delcourt and Delcourt, 1987). High pollen influx values may reflect increasing populations of trees; alternatively, the warm, dry climate of the mid-Holocene may have resulted in open-grown orchards that produced more pollen and nuts.

Humans may have helped to thin nut-bearing forest stands to encourage increased mast yield. Ethnobotanical studies confirm the regional utilization of mast nut resources by Archaic-period people from 11 500 to 3200 BP. Plant remains preserved in archaeological sites record increased use of hickory nuts during Middle Archaic time (8900 to 5700 BP) across mid-latitudes of the eastern United States (Ford, 1981).

Mast represented a highly nutritious food source. Nutmeats have 1.5 times the calories of beef (Talalay *et al.*, 1984), are high in fat, and contain nearly all essential amino acids (Gardner, 1997). Individual hickory or oak trees might provide nearly 20 kg of edible nutmeat; seven productive trees could sustain one person for a year (Gardner, 1997). Within the eastern deciduous forest, a watershed of 10-km radius can produce as much as 1.9 million kg (more than four million pounds) of acorns and 236 000 kg (more than 500 000 pounds) of hickory nuts per year. Mast nuts would have been collected during good mast years (Talalay *et al.*, 1984) and could have been stored up to three years. Once collected, mast nuts were probably hulled and heat-treated by parching, then dried to prevent spoilage (Petruso and Wickens, 1984). Archaeological evidence of nut processing includes stone-drilled depressions called bedrock mortars, or "hominy holes," containing organic residues dating from 4200 to 1850 BP, grinding pestles dating from 2400 BP, and ceramic boiling pots dating from 3200 to 670 BP. Even small storage pits, of 1 m depth and 1 m diameter, could contain 22 bushels of hickory nuts, some 140 kg of edible nutmeat satisfying one individual's daily needs for over 13 months (Gardner, 1997). By using a mortar and pestle to pulverize the nuts, then boiling them to concentrate the floating nutmeats and their oils, a person in an eight-hour session can process enough food to feed an adult for 16 days (Talalay *et al.*, 1984).

The Holocene distribution and ecological structure of oak–hickory–walnut forest communities, along with the autumn production of nuts, may have influenced behavioral and settlement patterns of Archaic people

(Gardner, 1997). Serving as a demographic bottleneck for Archaic foragers, late winter was the annual "hungry season" when white-tailed deer were lean and scarce, and when few other plant foods were available. The critical food value of mast resides in its storability, ensuring a predictable winter food supply. Success in collecting nuts would have been tied to a deliberate mast-exploitation strategy in which collecting trips would have been scheduled to coincide with nut fall and to focus on the most productive stands, ignoring less productive sites (Gardner, 1977).

The autumnal peak in nutfall for hickory nuts and walnuts coincides with the one or two weeks following the first killing frost (Talalay *et al.*, 1984). However, maximum nut concentration on the ground typically lasts no more than four weeks, as other woodland foragers such as squirrels and white-tailed deer compete for this ephemeral food opportunity. Similar patterns of nut yield occur for white oak and red oak species (Petruso and Wickens, 1984). Predictability of mast yield is also tied to the temporal pattern of cyclic nut production. Fluctuating between boom and bust yields, stands of hickory and oak regularly have abundant mast crops every two or three years. Timing of bumper mast years is typically offset between groups of red oak and white oak species, with nut yields by one oak group buffering shortfalls in the other. Extreme spring frosts may adversely impact pollination, fruit set, and nut harvest over the following one or two years (Gardner, 1997).

In addition to strategic collecting trips, woodland-dwelling people may have adopted a management strategy to enhance mast yield from known groves of highly productive trees (Munson, 1986). Archaic foragers could have thinned out brush and removed undesired or lesser-productive trees by ring-girdling their bark. The remaining nut-bearing trees would gain better canopy–crown access to sunlight. This limited but selective management would have created orchards of open-grown hickory, walnut, and oak trees. For example, selective girdling of individual hickory trees could have increased production of hickory nuts by up to 500 percent within a decade (Munson, 1986).

Witness-tree surveys conducted between AD 1715 and 1755 in southeastern Pennsylvania provide evidence that both hickory and walnut occurred in relatively high frequencies within a 7-km radius of late prehistoric Native American villages, where archaeological evidence indicated continuous human habitation extending back to Paleoindian times. Through time, a patchwork of temporally and spatially shifting settlements may have been sited specifically to exploit long-established and highly valued nut-resource orchards (Black and Abrams, 2001).

PANARCHICAL THRESHOLD FOR ARCHAIC FORAGERS

To account for archaeological evidence of a mid-Holocene increase in Middle Archaic human populations, Gardner (1997) championed a model of climate change and forest response that "released" the size of human populations, particularly after 8900 BP. The mid-Holocene interval across the Midwestern United States was characterized by a zonal atmospheric regime of strong westerly winds, with warm, dry, long growing seasons, and with lessening seasonal contrast and less frequent spring frosts than characterized the early-Holocene interval (Webb *et al.*, 1993). Mid-Holocene climate change drove regional compositional shifts in forest communities toward dominance of oak, hickory, and walnut between 8900 and 6800 BP (Delcourt and Delcourt, 1987). With fewer weather-related failures in spring pollination, mast production would have increased. With increased nut yield, Middle Archaic foragers specialized in hickory nuts as their most favored, nutritionally superior choice of food (Gardner, 1997).

As Archaic human populations grew, people dispersed to new satellite communities. Bands established a regionally extensive network of base camps along small tributary valleys as well as uplands. Such foraging populations, if engaged in widespread nut-tree management, would have been capable of altering the overall structure of forests. By thinning orchards and creating small forest gaps, Archaic people may have increased the forest-edge habitat for species of native herbs as well as white-tailed deer and other small game. This incipient forest fragmentation may have facilitated weedy herbs that became alternate food sources within the "Eastern Agricultural Complex" in Late Archaic times (Gardner, 1997).

In the late Holocene, the climate of eastern North America shifted to a north–south meridional circulation regime, with cooler, wetter conditions and more variable spring and fall "shoulder seasons" (Delcourt *et al.*, 2002). During this climatic transition between 4500 and 3200 BP, the shortening of the growing season and the increased frequency and severity of springtime killing frosts would have decreased the overall quantities and frequency of abundant nut harvests. Humans would have been forced to diversify their subsistence strategies beyond traditional mast exploitation strategies. Archaeological studies document the abandonment of major river base camps by late-Archaic people as they relocated to upland sites and small tributary valleys. This outmigration to the hinterlands is compatible with the interpretation that the population dispersed in response to less abundant and more unpredictable mast harvests.

Native weed seeds began to supplement declining mast yields. At the Archaic/Woodland transition *c.* 3200 BP, increased reliance upon ruderal plant species led to the beginnings of horticulture at upland rockshelter sites (Gardner, 1997).

INCIPIENT DOMESTICATION AND CURATION OF NUT-TREE ORCHARDS

The curation of "nut groves" in the vicinity of human habitations is an example of "incipient domestication" of selected species of nut-bearing trees (Rindos, 1989). Coevolutionary success would be measured in terms of both the effective dispersal of plant reproductive propagules and the learned subsistence behavior of humans whose own survival and reproductive success were thereby ensured. The interpretation of nut-tree curation, however, remains a speculation. One line of reasoning in support of this hypothesis compares nut-use to wood-use as a measure of exploitation. For example, nut-shell, seed, and wood–charcoal assemblages from the archaeological excavations of the American Bottom, near St. Louis, Missouri, show that hickory nuts and walnuts were important diet components through the Archaic and were supplemented with starchy and oily seeds of weedy plants after the Middle Woodland cultural period, particularly after 2100 BP. Although an excellent fuel for cooking and heating, the hard wood of hickory and walnut was uncommon in the corresponding wood–charcoal record. By inference, trees of hickory, walnut, and oak may have been valued more for their edible mast nutmeats than for their firewood (Rindos and Johannessen, 1991).

CONCLUSIONS: HUNTER–FORAGER RESOURCE STRATEGIES

Humans adapted to early-Holocene conditions by (1) diversifying their subsistence system; and (2) developing settlement networks that either were concentrated along increasingly sharpened biome boundaries such as the prairie/forest border, or that consisted of interconnected macro-band networks located along major waterways such as the Central Mississippi Valley, and rivers traversing the Fall Line and the Southern Atlantic Coastal Plain. Late Paleoindian and Early Archaic people practiced a logistically organized foraging strategy based upon establishing a semi-permanent residential base and a seasonal round of activities involving use of specialized field camps. Overall, the population sizes of Late Paleoindian (Folsom, Dalton) and Early Archaic hunter–gatherers were relatively small, and widely dispersed across the landscapes of eastern North America. Biome boundaries and watershed divides were buffer zones that separated macro-bands during most of each year. Late Paleoindian and Early Archaic people were more archaeologically visible than their Clovis and preClovis predecessors. Their lifeways as hunter–foragers (Figure 3.5) would have affected the patch structure and resource density locally around their seasonal settlements and base camps. Because of their small overall population sizes and their foraging strategies, however, they probably had minimal impact on regional landscape heterogeneity.

With the Holocene expansion of deciduous woodland, the emergence of canopy dominance of oak, hickory, and walnut created new opportunities for hunter–foragers, who developed behavioral strategies for exploiting geographically extensive resources of mast nuts. Success in collecting, preserving, and storing nuts was tied to a deliberate mast-exploitation strategy. Aboriginal bands had to schedule their collecting trips in order to coincide with the limited time of nutfall and to compete against other faunal foragers. Archaic people had to focus their efforts on the most productive stands and ignore less productive sites. Woodland-dwelling foragers may have adopted a management strategy of enhancing mast yield gleaned from these already rich resource patches. Prehistoric foragers may have improved microsite quality for their most prized nut trees by eliminating plant competition, by thinning out brush, and removing undesired or lesser-productive trees by ring-girdling trees. The remaining groves of curated, regularly-spaced and nut-bearing trees would have gained better canopy crown access to sunlight. In this way, prehistoric Native Americans may have directly influenced the population ecology of the nut trees as they increased the production of edible nut resources.

6
Community-level interactions

PREHISTORIC NATIVE AMERICANS AS AN INTERMEDIATE DISTURBANCE

By the late Holocene, after 4500 BP, Native Americans had demonstrable effects upon their ecological setting. Domestication of plants in riparian habitats resulted in cumulative effects on ecological relationships at the community level (Figure 3.4) as bottomland forests were cut and old-field communities developed. Late Archaic and Woodland people affected the proportions of native species in the vegetation surrounding their habitations by collecting fuelwood, selective use of fire, and gardening activities such as tillage, weeding, and establishing seed beds. These activities of prehistoric Native Americans constituted a form of intermediate disturbance that influenced biological diversity, both by altering species richness in the immediate surroundings of villages and rockshelters and by affecting the composition and distribution of plant communities along more extended ecological gradients from river bottoms to ridgetops.

THE RELATIONSHIP OF INTERMEDIATE DISTURBANCE TO BIOLOGICAL DIVERSITY

Several kinds of biological diversity can be measured directly. Alpha diversity, or within-habitat diversity, is determined by local species richness, measured by the total number of species found in a sample. Beta diversity, or between-habitat diversity, the rate of species turnover from one community to another along an ecological gradient, is measured by comparing the change in alpha diversity from one sample to the next. Gamma diversity, or landscape-level diversity, is the total number of species within a study area, where alpha diversity may change from one portion of an environmental gradient to another. Biological diversity may thus be measured as

(1) simple species richness; (2) the degree to which community composition changes across ecotones; or (3) the totality of the distribution of species abundances across a heterogeneous landscape mosaic (Whittaker, 1967, 1972, 1975; Gauch, 1982).

Studies of diversity patterns within tropical rain forests and coral reefs reveal an important relationship between species richness and disturbance (Connell, 1978). At very low and very high levels of disturbance, biological diversity is lowered, either because succession proceeds to a late stage in which only a few species dominate the community or because frequent and intense disruption continually sets back succession to early stages with few colonizers. At intermediate levels of disturbance, however, biological diversity reaches a maximum, especially along ecotones between communities where species from two communities may coexist. Disturbances may enhance species diversity by (1) lowering the dominance of one or a few species, and freeing up resources for other less competitive species (Paine, 1966; Connell, 1978; Lubchenco, 1978); or (2) increasing environmental heterogeneity, and thus providing a basis for specialization and resource partitioning (Denslow, 1980, 1985; Tilman, 1982). Disturbances thus affect not only the relative abundances of species but also the spatial and temporal heterogeneity of ecosystems (Hunter *et al.*, 1988). An intermediate level of disturbance helps to preserve the patch structure and competitive conditions upon which some species depend. The composition of plant colonizers on a newly created disturbance patch is a result of germination of seeds from the dormant seed bank, resprouting of perennials from belowground rhizomes or root systems, and the seed rain reaching the patch from plant populations in nearby patches. In species-rich communities, strong variation both seasonally and spatially is typical among species colonizing disturbance patches (Denslow, 1985).

ORIGIN OF OLD-FIELD COMMUNITIES

Disturbances that open up light gaps within otherwise closed forest may also encourage the invasion of weedy species (Low, 1999; Simberloff and Von Holle, 1999). Activities of prehistoric Native Americans that disturbed ever-increasing patches thus would have been a mechanism for the spread of native weeds.

Information on seed dispersal and native habitats for 13 indigenous plant species that today are important constituents of old-field communities growing on abandoned agricultural land shows that in addition to growing in old-fields, most of the species of grasses, forbs, and shrubs occurred in "marginal" and "persistent open" habitats (Marks, 1983). Naturally occurring open environments in which closed forest cannot develop, such as rock outcrops or point bars along stream beds, were likely habitats in which the native species of plants today characteristic of old-field communities lived prior to European American land clearance. Most of the species included in Marks' study had relatively poor seed dispersal, and thus he considered that these plants would not have readily colonized forest-canopy gaps on their own; rather, he considered river valleys as "corridors of relatively continuous marginal habitat" suitable for long-term persistence of patchy species populations (Marks, 1983). Marks further discounted the hypothesis that "native field plants evolved on agricultural land abandoned by Indians far back in time," arguing that one would still have to account for the whereabouts of the plants before that time.

The analysis of Marks (1983) included several plants that are known from ethnobotanical studies (Cowan *et al.*, 1981; Smith, 1992; Gremillion, 1993b) to have been part of the indigenous flora directly used by prehistoric Native Americans. These plants include ragweed (*Ambrosia*), bluestem grass (*Andropogon*), strawberry (*Fragaria virginiana*), and sumac (*Rhus typhina*). Other included taxa, such as red cedar (*Juniperus virginiana*), have been shown to increase in representation in ethnobotanical samples from Late Archaic to Woodland times, especially in archaeological sites situated on low stream terraces (Chapman *et al.*, 1982; Bareis and Porter, 1984; Delcourt *et al.*, 1986). Even hearths within rockshelters located along ridgetops situated high above floodplains, and that were subject to Indian occupation and burning, contain charred seeds of ragweed and little bluestem (*Andropogon scoparius*), in addition to seeds of other plants characteristic of the Eastern Agricultural Complex (Gremillion, 1993b; Delcourt *et al.*, 1998; Figure 6.1).

Indigenous species of plants that were restricted to bluffs, rock outcrops, localized cedar glades, and stream beds in the early and mid-Holocene thus were spread beyond those habitats in the late Holocene, between 4500 and 3000 BP. They were distributed to newly expanded niches not by their own seed dispersal but by the intervention of prehistoric Native Americans, who carried seeds with them on seasonal migrations between floodplains and ridgetops. Through their activities, Late Archaic and Woodland people represented an intermediate disturbance that allowed expansion of the niche breadth of the plant species.

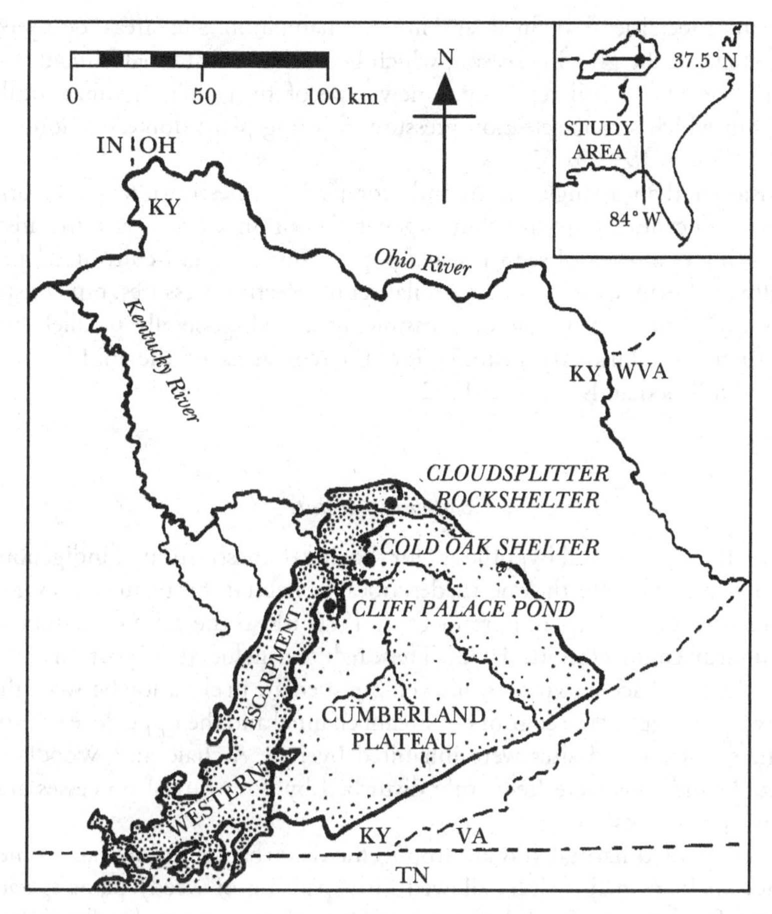

Figure 6.1 Environmental setting of paleoecological and archaeological sites in the Daniel Boone National Forest, located on the Northern Cumberland Plateau and Western Escarpment region of eastern Kentucky (modified from Delcourt *et al.*, 1998).

INTERMEDIATE DISTURBANCE AND THE EASTERN AGRICULTURAL COMPLEX

Human disturbance was variable both spatially and temporally throughout the region of the mid-continent of North America in which the Eastern Agricultural Complex was domesticated (Smith, 1978, 1992). Meander belts along large river systems, such as the Mississippi River and its major tributaries, were the sites for development of Late Archaic and Woodland "domestilocalities," defined as "human habitation site areas occupied throughout the growing season, which because of substantial human disturbance of the soil represent a new type of river valley habitat locale within which strong selection pressures favoring plant domestication occur" (Smith, 1992, p. 62).

Rather than a single "heartland" for plant domestication, prehistoric domestilocalities were distributed generally within a broad riverine area, resulting in a series of semi-isolated populations of "quasi-cultivated" and cultivated plants, each under a similar set of selective pressures, but constituting a "complex mosaic of occasionally linked, generally parallel, but distinct coevolutionary histories for different areas of the mid-latitude Eastern Woodlands" (Smith, 1992).

Floodplain habitats

The "floodplain weed hypothesis" for the first domestication of indigenous plants in eastern North America describes the habitats for plant species that became economically important from 4500 BP to the time of European American contact (Smith, 1992). These habitats included the portion of the floodplain adjacent to the stream, yet intermediate in elevation between the low-water stage at the edge of the stream channel and the upper levee crests. These bottomland sites were inhabited by Late Archaic and Woodland people and thus were frequently disturbed by both fluvial processes and human activities.

Augmented natural and anthropogenic disturbance regimes along meander belts of major rivers allowed for expansion of weedy plant species out of more restricted habitats into gaps opened in the floodplain forest canopy, leading to microevolutionary changes and areal expansion into newly available niches (Smith, 1992). These new niches became filled by early-successional plant species and marked the development of incipient old-fields, known to historic southeastern Indians as "tallahassees" (Chapman *et al.*, 1982; Delcourt *et al.*, 1986; Delcourt, 1987).

Plant–ecological analyses of a number of floodplain sites in Illinois (Asch and Asch, 1978; Munson, 1984) and throughout the Midwestern United States (Smith, 1989, 1992) have led to an understanding of primary natural and anthropogenic habitats for two of the principal species, goosefoot and marsh elder, cultivated in prehistoric times as part of the Eastern Agricultural Complex.

Goosefoot, Chenopodium berlandieri

Goosefoot exhibited a progressive thinning in seed coat through its archaeological record (Figure 4.2). This important prehistoric domesticated plant is an early successional plant species the primary habitat of which is along river valley floodplains of large meandering rivers such as the Mississippi. *Chenopodium berlandieri* occurs in a number of different kinds of environments characterized by disturbed soil. Natural habitats include (1) the shady understory of black willow stands along sand banks of river margins where annual scouring by floodwater is followed by deposition of sandy alluvium; (2) sandy beaches along river margins left bare by receding water in summer; and (3) eroding river and terrace banks. Plants growing in a shady understory are tall and produce relatively few seeds.

In anthropogenic habitats, including fields and gardens along meander belts and active main channels of river valleys, extensive stands of goosefoot are characterized by multiple branching of densely grown plants. Smith (1992) concluded that poorly weeded, overgrown fields and gardens on natural levees would have been the primary sites for the most productive stands of goosefoot in prehistoric times, similar to the sand-bank gardens observed by the Spanish at first contact with the Natchez Indians of southwestern Mississippi.

Marsh elder, or sumpweed, Iva annua

From the archaeological record of its use, marsh elder is a species that showed increased achene size during the transition from foraging to farming (Figure 4.2). Marsh elder has a habitat range that is defined by the reach of seasonal floodwaters. Dispersal of its achenes, which ripen and fall from the plant in autumn, occurs through winter and spring overbank flooding. Asch and Asch (1978) described *Iva* as an edge species that becomes well established on sites located primarily between permanently wet and well-drained soils.

Stands of *Iva* grow as large as 1000 m^2, and include as many as 300 000 individual plants in their optimal habitat zone of seasonally inundated soils in slackwater areas behind levee crests. This riparian habitat is characterized by low topographic relief, poor drainage, and long hydroperiods of inundation. Humans may disperse *Iva* to upland habitats, but plant populations observed around upland house sites consist of only a few plants, not extensive stands (Smith, 1992).

LONG-TERM VEGETATION CHANGE AND INTERMEDIATE HUMAN DISTURBANCE ON THE CUMBERLAND PLATEAU OF KENTUCKY

Native Americans have long been attracted to the shelter and valued resources of food and chert provided by the deeply incised river gorges and flat-topped mesas of Eastern Kentucky's Cumberland Plateau (Figure 6.1). There, hundreds of rockshelters and shallow caves (Smalley, 1986) contain layered deposits that preserve an archaeological record of human occupations extending back through the Early Paleoindian period to 14 000 BP (Ison, 1991, 1996; Gremillion, 1999). Excavated archaeological sites such as Cloudsplitter Rockshelter (Cowan *et al.* 1981; Cowan, 1985a, b, 1997), Cold Oak Shelter (Ison, 1988; Gremillion and Ison, 1992; Gremillion, 1993a, b, 1996, 1997) and Cliff Palace Cave (Delcourt *et al.*, 1998) contain plant remains selected and used by prehistoric people (Figures 6.1, 6.2, 6.3).

The fine-scale record of prehistoric human impact can be demonstrated in the integration of three kinds of evidence: (1) the archaeological chronology for human use of rockshelters; (2) archaeological documentation of the deliberate collection and cultivation of plants; and (3) the paleoecological record of local vegetation and wildfire history. Archaic and Woodland people inhabited sandstone ledge rockshelters that offered ready access to upland pools and springhead seeps that provided fresh drinking water.

Prehistoric human occupation in the Daniel Boone National Forest

Radiocarbon analyses of organic matter from hearths, and thermoluminescence dates of sediments, provide 110 absolute dates of human occupation of rockshelters within the Daniel Boone National Forest (Ison, 1996; Cowan *et al.* 1981; Cowan, 1997; Gremillion, 1997, 1999). Rockshelters were used intermittently throughout the Paleoindian, Early Archaic, and Middle Archaic cultural periods (Figure 6.2). In Late Archaic through Late Woodland times, rockshelters were occupied regularly, as evidenced by the large number of ethnobotanical samples dating from 3200 to 1000 BP. Absolute dates abruptly diminish in both numbers and consistency after 1000 BP, supporting the interpretation that the Cumberland Plateau region was largely abandoned by that time.

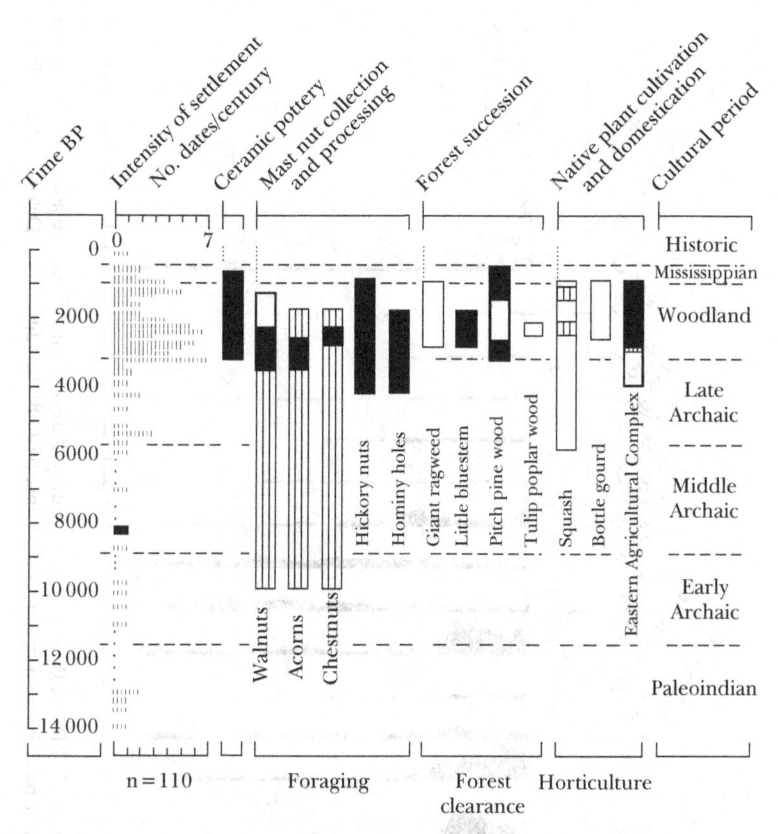

Figure 6.2 Hunter–forager and forager–horticulturalist ecosystems of the Northern Cumberland Plateau, eastern Kentucky. Chronology of aboriginal settlement intensity is expressed as number of absolute dates ($n = 110$) per century. Ethnobotanical evidence includes data relating to human activities, including plant taxa and archaeological features that are indicators of mast nut collection and processing, forest clearance and succession, and native plant cultivation and domestication. On the vertical bars, open bar segments represent rare occurrence in ethnobotanical samples, vertical stripes denote common occurrence, and black bar segments indicate abundant occurrence.

Based upon the changing character of archaeological features and hearths preserved at Cloudsplitter Cave, a demographic shift took place in Late Woodland times, 2000 to 1000 BP (Cowan, 1997, p. 84):

Late Woodland residency of the Plateau was restricted to short-term visits in the fall, perhaps to hunt selected animal prey species, followed by winter abandonment. Late Woodland populations seem to have spent the winter months in large

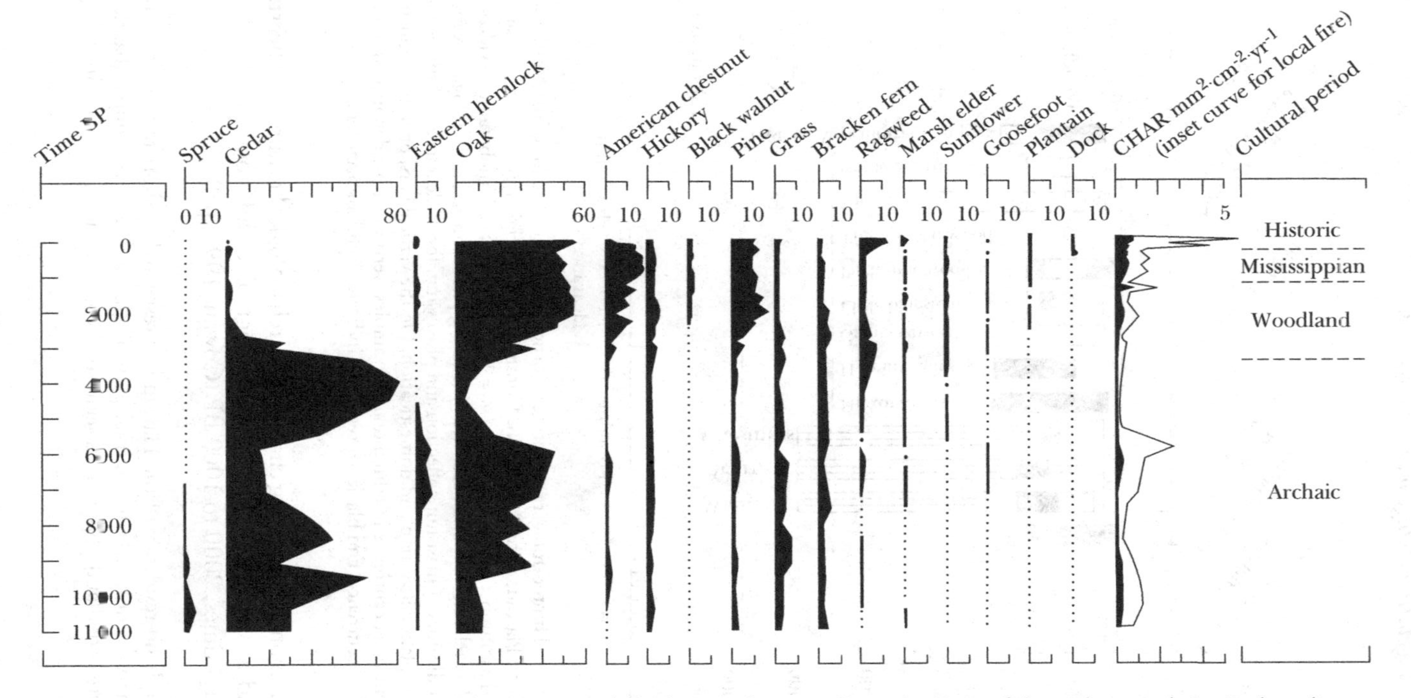

Figure 6.3 Holocene vegetation history and fire regime at Cliff Palace Pond, eastern Kentucky. Rates of charcoal accumulation in the sediment (CHAR) indicate increased prominence of local fires (the black inset for large particles) since 3200 BP (modified from Delcourt *et al.*, 1998).

nucleated settlements located in regions outside the Cumberland Plateau. During the midspring, small family groups probably dispersed from these villages, some of which returned to Plateau rockshelters in order to plant small gardens that could be harvested during a return trip in the fall. After the gardens were planted, the families went to the larger settlement, spending the summer months harvesting aquatic resources and mammalian resources, before removing to the Plateau for fall hunting.

Ethnobotanical and paleoecological evidence for prehistoric community-level interactions

The wealth of radiocarbon-dated ethnobotanical assemblages from archaeological sites on the Cumberland Plateau reflect aboriginal utilization of plant resources for firewood and for storable plant food (Figure 6.2). The archaeological record of plants from rockshelters documents substantive forest fragmentation at about 3200 BP, as garden plots were cleared within nearby forests. The creation, use, then abandonment of garden plots provided disturbed sites for colonization by weedy herbs including ragweed and little bluestem, as well as early-successional trees including tulip poplar (*Liriodendron tulipifera*) and pitch pine (*Pinus rigida*). After 1000 BP, Native Americans no longer used the rockshelters on the Cumberland Plateau for cultivation of species of the Eastern Agricultural Complex or for harvesting and processing mast nuts (Fig. 6.2).

Paleoecological evidence for community-level effects of prehistoric human activities on the Cumberland Plateau is available from sediments analyzed from a 5 m × 15 m woodland hollow (Delcourt *et al.*, 1998). Cliff Palace Pond is perched on the crest of a narrow ridge of plateau called Keener Point. With a small (0.5 ha) watershed, the pond collects airborne pollen derived primarily from a 1-km radius surrounding the immediate ridgetop, representing the vegetation growing on upper slopes of nearby steep-sided ravines (Figure 6.3). The Cliff Palace complex of rockshelters, located 60 meters west of the pond and mid-way down the hill slope, preserves a local record of diagnostic stone artifacts that document human presence extending back to at least 9000 BP. Human activities on Keener Point encompassed both the rockshelter and the pond site. The two areas are linked by a series of prehistoric stair steps that were hand-carved into the sandstone cliff wall. Taken together, the archaeological and paleoecological records from Cliff Palace rockshelter and pond give evidence of the role of prehistoric Native Americans in plant community dynamics through the Holocene.

Cliff Palace Pond sediments contain a record of vegetation change for the past 10 700 calendar years. In the early Holocene, cool-temperate forests of northern white cedar (*Thuja occidentalis*), red spruce (*Picea rubens*), and oak were replaced by mixed mesophytic forests of oak and other temperate hardwoods from 8100 to 5500 BP. Mid-Holocene forests were relatively species-poor and were composed of fire-intolerant species until about 3200 BP (Figure 6.3). After 3200 BP, fire-tolerant, mixed deciduous–evergreen forest established and was dominated by oak, chestnut, hickory, walnut, and pitch pine.

Human presence in the rockshelters flanking Cliff Palace Pond is documented with ethnobotanical remains that have been dated at 3650, 3200, 2850, and 1600 BP (Delcourt *et al.*, 1998; Carmean and Sharp, 1998). At the time of the Archaic/Woodland cultural transition, peaks in herbaceous pollen of ragweed, grass (Poaceae), plantain (*Plantago*), and spores of bracken fern (*Pteridium aquilinum*) in Cliff Palace Pond sediments indicate the expansion of open patches in the landscape. The continuous occurrence in the paleoecological record of pollen of marsh elder, sunflower, and goosefoot after 3000 BP is evidence that Eastern Agricultural Complex species were grown by Native Americans in garden plots on upper slopes, adjacent rockshelters, and possibly also near Cliff Palace Pond on Keener Point (Delcourt *et al.*, 1998).

Charcoal particles preserved, along with pollen grains, in the Holocene pond sediments yield evidence of past fire regimes. The largest particles of charcoal (>5.0 grid squares in cross-sectional area and >50 μm in length) provide conservative evidence for local fires near the pond (Clark, 1988a, b). Rates of charcoal accumulation in the sediment (CHAR) provide a quantitative index of changes in the magnitude of prehistoric fires through time. Both the proportion of large particles and overall charcoal influx show marked increases corresponding with the Woodland-age occupation of the Cliff Palace Cave (Figure 6.3)

We interpret the strong coincidence in time of prehistoric human occupation in local rockshelters, ethnobotanical remains of domesticated native plants, and increases in local fires shown by the charcoal record from Cliff Palace Pond as evidence for a cause-and-effect relationship between Native American activities and changes in forest composition during the past 3200 calendar years. Human-set fires would not only have helped people to establish garden plots, but they also would have facilitated regeneration of oak and pine and thereby promoted a change in plant communities to include fire-tolerant species (Figures 6.2, 6.3).

Cliff Palace is only one of many rockshelter sites near which prehistoric Native Americans of Late Archaic and Woodland cultural periods concentrated their farming activities (Cowan 1985a, b; Ison 1988, 1991). Human use of fire in clearing garden plots would have affected the forest vegetation in many locations on upper slopes and ridgetops immediately surrounding their habitations throughout the Cumberland Plateau region.

At a site such as Cliff Palace Cave, the complex of nearby rockshelters may have been home to several families at a time (Sharp 1997; Carmean and Sharp, 1998). The extent of forest clearing needed to provide three families with sufficient calories from the kinds of seeds and fruits of crops typically found in Late Archaic and Early Woodland hearths for even one winter month is estimated at 10 to 40 m^2 (Cowan 1985a). Clearance of a forest gap of sufficient size to feed three families over the winter months would result in an open area on mid-slopes near the rock shelter of nearly 400 m^2, large enough for succession of native hardwood species, including tulip poplar, which requires light gaps of this spatial extent for regeneration (Runkle 1985). Insect-pollinated tulip poplar is represented consistently both in ethnobotanical remains (Figure 6.2) and in pollen assemblages in the interval associated with Early Woodland people's cultivation of plants of the Eastern Agricultural Complex.

The major expansion of culturally augmented wildfire after 3000 BP favored a changeover in forest composition and structure. Populations of fire-adapted oaks, chestnuts, hickories, and walnuts rose to canopy dominance, as these mast-bearing hardwoods replaced fire-intolerant species. The result of anthropogenic fire was a fine-grained patchwork of vegetation on upper hillslopes and ridgetops that included prehistoric garden plots, open patches with mixed crops of domesticated species, abandoned Indian old-fields reverting back into early-successional grassland barrens, thickets of shrubs, and even-aged stands of pitch pine or tulip poplar trees. At the base of the ravine gorges, however, moist, shady habitats on talus-strewn slopes or calcareous bedrock may have continued to provide refuges for fire-intolerant cove, or mixed mesophytic, forest communities.

The changeover from fire-intolerant to fire-tolerant forest on upper hillslopes took place during the late Holocene, when the regional climate was becoming cooler and wetter. This change in forest composition coincided, however, with the time of most extensive human use of rockshelters (Figure 6.2). Through their use of fire, Native Americans profoundly influenced the composition of forests in a manner opposite to that expected from the prevailing climate. By Late Woodland times, however, a demographic shift resulting in the use of rockshelters only briefly during fall and spring visits decreased the amount of human disturbance. This resulted in expansion of pine stands on abandoned old-fields, documented in the local ethnobotanical record after 1500 BP (Figure 6.2).

On the Cumberland Plateau of Kentucky, between 3200 and 1000 BP aboriginal use of wildfire for landscape management generated an intermediate disturbance regime that enhanced landscape-level heterogeneity. Through use of fire, prehistoric Native Americans effected a community-level shift in forest composition to oak–chestnut–hickory forests in the uplands. We also infer that the overall ecological gradient was broadened to include mesic forest communities on fire-protected, lower slopes, with fire-adapted species invading fire-prone upper slopes and ridgetops. Alpha diversity increased locally, beta diversity increased along topographic gradients from ridgetops to valley bottoms, and gamma diversity increased during this time of finer-grained vegetation patchwork and increased landscape heterogeneity.

ECOLOGICAL RESISTANCE TO INVASION

Prehistoric Native Americans and plants interacted in a synergistic manner (Delcourt *et al.*, 1993; Simberloff and Von Holle, 1999; Richardson *et al.*, 2000), developing mutualistic relationships in which the dispersal of seeds, local abundance of plants, and degree of aggregation of plants were all enhanced as humans adopted behaviors that included tending, tilling, and selective harvesting of plants for food and fiber.

Synergistic influences, where the impact of several species taken together is greater than the sum of the impacts of the individual species, would have affected the ecosystem as a whole (Simberloff and Von Holle, 1999; Richardson *et al.*, 2000). For example, in human-disturbed habitats, certain plants such as locust trees (*Robinia, Gleditsia*), alder (*Alnus*), wax-myrtle (*Myrica*), ground-nut (*Apios*), and other legumes that are disturbance-adapted are also capable of modifying the nutrient status of the soil through symbiosis with nitrogen-fixing bacteria, making more suitable sites for invasion by other species favored by humans for food.

Indirect effects of humans would have modified the habitat around domestilocalities (Low, 1999). For example, frequently used trails and footpaths would have been sites suitable for invasion and establishment of plants used as potherbs (plantain [*Plantago*], pokeweed [*Phytolacca*]), fiber plants (nettle [*Laportea, Urtica*], milkweed [*Asclepius*], Indian hemp [*Apocynum*]), and dye plants (bedstraw [*Galium*]).

Relationship of human disturbance to changes in ecological resistance

Over time, the interplay of humans and environment on a local scale (Delcourt and Delcourt, 1997) would have led to incipient old-field succession, perhaps by multiple relay invasions (Egler, 1977) of riparian and marginal- (edge-) habitat species. Under a new, human-mediated intermediate disturbance regime, with changes in disturbance frequency, openness of local landscapes, and changes in soil conditions, prehistoric riparian communities may have become open to local invasions of plant species spreading beyond their native habitats in marginal environments. Human activities that facilitated such expansions in niche breadth may have lowered the "ecological resistance" to invasion (Lodge, 1993a, b; Von Holle *et al.*, 2003). Only at the end of the Mississippian cultural period, after broad-scale forest clearance, development of large villages, widespread agriculture on low river terraces, and abandonment of Indian old-fields, did secondary succession result in closed vegetation that was resistant to further invasion.

Canebrakes as an example of changing ecological resistance

With moderately intense disturbance, dense canebrakes form on levee crests, observed in early historic times as wide as 1.5 km along the Mississippi River and its major tributaries (Platt and Brantley, 1997). Many of the canebrakes described by European settlers and explorers probably originated by human facilitation. Prehistoric Native Americans maintained stands of cane for basketry and housing construction (Chapman, 1994).

Dense, impenetrable canebrakes formed, however, not as a consequence of intermediate levels of disturbance around Indian village sites, but rather as a result of rapid invasion of fallow land following widespread land abandonment in the sixteenth and seventeenth centuries (Hudson, 1997; Platt and Brantley, 1997). These historic canebrakes were relatively species-poor habitats, and would have become highly resistant to further invasion by other native plants because of their dense shade and equally dense underground rhizome system. Native species of plants previously used for either food or fiber would not have been competitive invaders of dense canebrake vegetation that was not periodically harvested for domestic use. Only with European American clearance of the canebrakes for sugar cane production and with intense grazing pressure from domestic cattle were these habitats subsequently opened up to invasion by a wide variety of herbaceous and woody plant species (Hamel and Buckner, 1998). Successful invasive plants of riparian habitats in the past 200 calendar years are largely exotic species introduced to the New World from the Old World (Crosby, 1986).

LEVELS OF NATIVE BIODIVERSITY IN EASTERN NORTH AMERICA

Ethnohistoric data describing early historic landscapes of the southeastern United States support the interpretation that at least eight types of patches were anthropogenically maintained: (1) hunting camps; (2) fields and gardens; (3) residential sites; (4) edge areas and meadows; (5) old-fields; (6) parklands and orchards; (7) wetlands, swamps, and marshes; and (8) waterways (Hammett, 1992, 1997, 2000). Localized thinning and clearing of vegetation near settlements, in part through the deliberate use of prescribed burning, may have been crucial to maintaining open landscape patches for hunting camps, meadows, parklands, and nut-tree orchards in prehistoric times (Hammett, 2000).

In late prehistoric times, the influence of aboriginal land clearance in the lower Southeast was in general confined to the immediate vicinity of villages concentrated along major river valleys; large portions of the landscape on upland drainage divides, at some distance from human habitations, were vegetated by natural forest communities (Fritz, 2000). Large agricultural fields that remained open for long periods of time were, however, only one of the kinds of anthropogenically altered ecosystem types in prehistoric eastern North America, representing an extreme endpoint in a gradient of human influence that ran the gamut from little discernible impact to land clearance and cultivation. Fritz (2000) maintained that, in most instances, prehistoric Native Americans were an ecological influence operative at a local scale, with intermediate environmental impacts.

CONCLUSIONS: HUMANS AS AN INTERMEDIATE DISTURBANCE REGIME

Beginning in the Late Archaic and Early Woodland cultural periods, prehistoric Native Americans began to have discernible impacts on native vegetation at the community level. Settlements in villages or domestilocalities along meander trains of major river systems in eastern North America were sites not only for domestication of indigenous plants but also for succession of old-field weed communities. Changes in niche breadth of riparian species facilitated by human disturbance near prehistoric habitations resulted in incipient old-field succession as newly opened habitats were invaded by aggressive weedy plant species.

At rockshelter sites from the Ozark Highlands to the Appalachian Mountains, Late Archaic and Woodland people grew indigenous food plants in garden plots that, with girdling of trees and localized use of fire, opened up gaps in the forest canopy allowing for a patchwork of forest succession. On the Cumberland Plateau of Kentucky, human-set fires burned upper slopes, broadening the ecological gradient to include both fire-intolerant and fire-adapted plant communities.

Effects on biological diversity occurred through increased plant competition for sunlight, facilitated by anthropogenic opening of disturbance patches such as light gaps in the forest canopy. Increased competition for below-ground water and nutrients would have been aided by manipulating the soil, increasing aeration and preparing a seed bed for domesticated plants, as well as by thinning of underground stems and roots, and selective encouragement of plants with symbiotic nitrogen-fixing bacteria adaptations.

Taken together, human activities characteristic of Late Archaic and Woodland period lifeways constituted a sustained, intermediate disturbance regime that resulted in regeneration of biologically diverse landscape patches in a wide variety of ecological settings.

7
Landscape-level interactions

ANTHROPOGENIC DISTURBANCE, FOREST FRAGMENTATION, AND LANDSCAPE STABILITY

Along the continuum from incipient domestication of plants, to slash-and-burn cultivation of village garden plots, to development of extensive agricultural fields, several kinds of landscape thresholds may be crossed. These thresholds involve geomorphic processes, hydrologic changes, and habitat continuity. Anthropogenic disturbance, if severe and extensive enough, can result in over-exploitation of resources leading to forest fragmentation, soil degradation, and deterioration of the environment, with potentially irreversible effects (Figure 3.4). According to Redman (1999, p. xi): "The archaeological record is populated with thousands of communities [worldwide] that for varying lengths of time maintained a balance with their environment, yet over time, virtually all developed practices that degraded their surroundings and undermined their continued existence."

Native American disturbance and landscape change: Crawford Lake, Ontario

High-resolution pollen, plant macrofossil, and charcoal records from southern Ontario give definitive evidence that Native Americans impacted the vegetation during the time interval from 650 to 250 BP, before the time of European American settlement of the region (McAndrews and Boyko-Diakonow, 1989; Byrne and Finlayson, 1998). The evidence is derived from annually laminated sediments from Crawford Lake, located 65 km southwest of Toronto, along the Niagara Escarpment, within the northern conifer–mixed hardwoods forest region (Figure 7.1).

Because the Crawford Lake basin is relatively small (2.4 ha surface area), deep (24 m water depth), and steep-sided, the lake is meromictic and has deposited sediments in annual varves. The sediment was sampled with resolution ranging from 10-year intervals between 950 and 160 BP

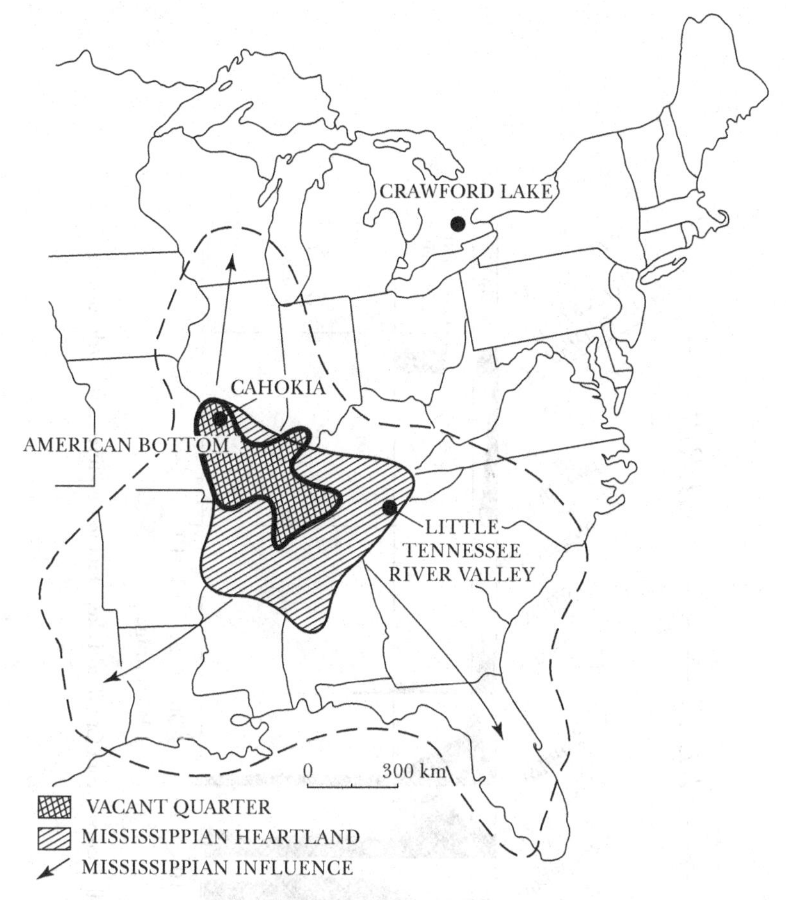

Figure 7.1 Archaeological sites documenting evidence for agrarian ecosystems during the Mississippian cultural period at Crawford Lake, southern Ontario, the Little Tennessee River Valley, eastern Tennessee, and the Cahokia metropolis in the American Bottom, southern Illinois. Agriculturalist human ecosystems predominated in the Mississippian heartland beginning 1000 BP. This core area was abandoned 500 BP, leaving a "vacant quarter" that underwent plant succession to secondary forests and wetlands (modified from Williams, 1982).

to 5-year intervals from 160 BP to the present (McAndrews and Boyko-Diakonow, 1989). Pollen analysis revealed a dramatic change in forest composition between 600 and 100 BP (Figure 7.2). By around 600 BP, forests surrounding Crawford Lake were largely composed of deciduous trees, including American beech, sugar maple (*Acer saccharum*), elm (*Ulmus*), basswood (*Tilia*), hickory, butternut (*Juglans cinerea*), birch (*Betula*), and

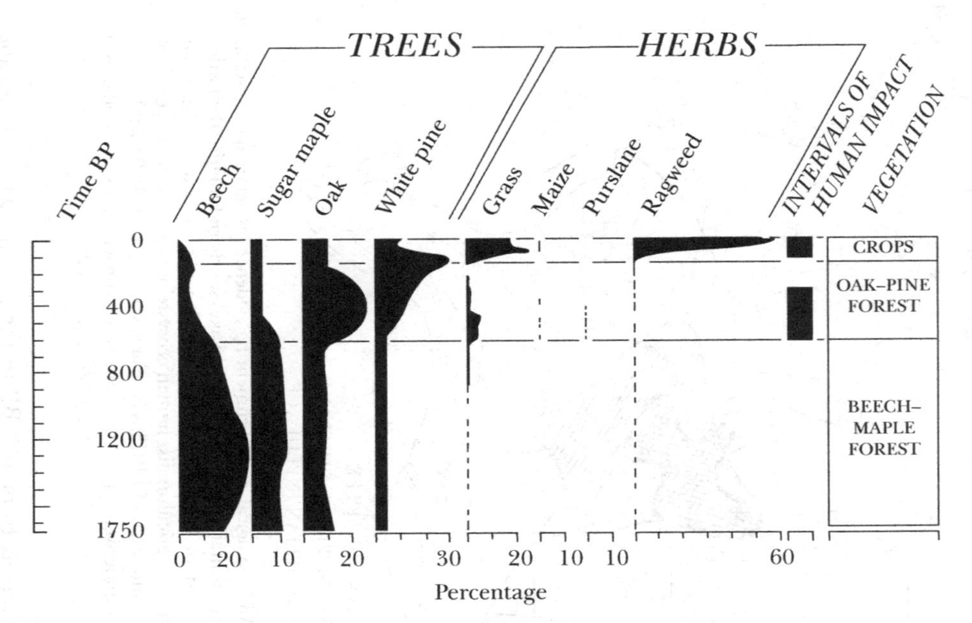

Figure 7.2 Selected pollen curves from Crawford Lake, Ontario (modified from Delcourt, 1987).

aspen (*Populus*), with only minor amounts of northern white cedar, white pine (*Pinus strobus*), and hemlock (*Tsuga canadensis*). American beech, sugar maple, and basswood all declined in importance and were replaced by oak and white pine after 600 BP. After reaching a peak in representation at 300 BP, oak was replaced by white pine around 100 BP. European American settlers encountered forests dominated by white pine, which became a valuable timber tree (Whitney, 1994).

Pollen evidence of Iroquoian agriculture was found in the sediment cores from Crawford Lake, dating from 650 to 280 BP (Figure 7.2). This evidence was the basis for locating and excavating a prehistoric Iroquoian village on the banks of the lake. Increases in pollen of grasses, along with the presence of both pollen from maize and the herbaceous ruderal, purslane, occurred in sediments representing the time of changeover from beech–maple forest to oak–pine forest (Byrne and McAndrews, 1975), and were interpreted as the result of aboriginal agriculture (McAndrews and Boyko-Diakonow, 1989).

Little Ice Age climate change or prehistoric human use of fire?

The change in forest composition from hardwoods to pine at Crawford Lake may have resulted, in part, from climate-cooling associated with the "Little Ice Age" episode of climate-cooling, which may have caused a southward shift in the range of white pine trees during the interval between 500 and 100 BP. With a 1–2 °C decrease in mean annual temperature, beech trees may have become less competitive at the latitude of southern Ontario. With population dieback of beech, canopy gaps would have been colonized by white pine and other pioneer tree species (Campbell and Campbell, 1992, 1994).

Either climate change or forest clearance by Native Americans could have altered the competitive advantages of taxa, resulting in changes in forest composition (Campbell and McAndrews, 1993, 1995). Analysis of the fossil charcoal content of Crawford Lake sediments showed that from 1750 to 590 BP, accumulation rates of particles >50 μm were less than $1\ mm^2 \cdot cm^{-2} \cdot yr^{-1}$ (Clark and Royall, 1995). Charcoal accumulation rates rose during the period of Iroquoian occupation, then declined until the beginning of the twentieth century. Clark and Royall (1995) concluded that an anthropogenic disturbance regime, the intentional burning of the forest by the Iroquoians, therefore was the major factor in the change in forest composition.

Phases of forest clearance

During the interval of Iroquoian occupation, eight phases of forest clearance have been documented using both pollen assemblages and influx of charcoal particles (Byrne and Finlayson, 1998). Evidence of local cultivation of plants included pollen from maize and cucurbits, pollen and seeds of sunflower, and pollen and seeds of purslane. Weeds that would have grown in cornfields were represented by carpetweed and goosefoot, and plants growing on abandoned Indian old-fields included grasses, asters, and bracken fern. Elevated charcoal influx values were associated with the times of most intense cultivation of plants.

The land-clearance phases reconstructed from the Crawford Lake sediment core were correlated with Iroquoian occupations documented from within 5 km of the lake. The occupation sites represent shifting villages of limited duration and changing proximity to the lake site, with relatively low populations ranging from 320 to 3100 people per village, and with individual villages covering between 0.7 and 5.5 ha. For example, the Crawford Lake archaeological site was first occupied between 515 and 491 BP. Located on a drumlin ridge 200 m northwest of the lake, the village site was 0.7 ha in area, home for an estimated 204 to 228 people. Lake sediments deposited during the time this site was occupied contain pollen from maize, purslane, cucurbits and grasses, and high peaks of charcoal, indicating that the forest adjacent to the lake had been cleared, burned, and farmed (Byrne and Finlayson, 1998).

Detailed analysis of the correspondence between timing of prehistoric Indian occupation, pollen and seed indicators of human activities, and charcoal evidence of human-caused fires supports the interpretation that even "relatively small populations of shifting agriculturalists can, over a period of several hundred years, have a major impact on the environment," in this case, converting the upland forest from late-successional deciduous forest to a patchwork of early-successional red oak and pine forest and old-fields (Byrne and Finlayson, 1998, p. 107).

Dey and Guyette (2000) extended these results for prehistoric human impact upon Ontario landscapes by examining six stands of northern red oak (*Quercus rubra*) and their fire histories, extending back to 300 BP. Using tree-ring sequences, fire scars were dated from stumps and living trees of red oak, white pine, and red pine (*Pinus resinosa*). For the interval from 300 to 100 BP, these fire histories were compared with early historic estimates of human population densities. At each site, recurrence intervals for fire events were shorter during times of increases in human population density. A shifting mosaic of cultural settlement, abandonment, and reoccupation may have facilitated regeneration of red oak trees by increasing the frequency of ground fires that favored germination of acorns (Dey and Guyette, 2000).

These southern Ontario studies demonstrate that even the activities of shifting agriculturalists can have cumulative effects. PreIroquoian forests were relatively homogeneous, late-successional northern hardwood forest in which the return interval for natural disturbance (wildfire or windthrow) would have been relatively long in comparison to the lifetimes of the trees. This type of natural disturbance regime favors the development of all-aged stands that remain in a steady state through regeneration in local canopy gaps (Bormann and Likens, 1979; Turner *et al.*, 1993).

The swidden agriculture activities of the Iroquois, however, changed the disturbance regime. With intermediate levels of disturbance, prehistoric people introduced a shortened recurrence interval for fire and cleared larger patches of the landscape than occurred with individual tree falls. With a disturbance interval shorter than the recovery time of the forest ecosystem, and with disturbance opening up larger and ever-expanding patches of open landscape, the vegetation shifted to an early-successional mosaic (Turner *et al.*, 1993). It became a more heterogeneous patchwork of oak–pine forest, mixed hardwoods, and Indian old-fields (Byrne and Finlayson, 1998). As this occurred, the ecotone between forested and nonforested landscape patches would have increased in breadth and length, in turn influencing a change in the fauna from predominantly forest-interior species to those favored by the edge effect of interspersed forests and fields.

Iroquois villages near Crawford Lake were abandoned by the time European American pioneers began to settle on the land. European Americans, however, mistakenly viewed the large groves of white pine as a "virgin" forest. Paradoxically, in southern Ontario, even though the impacts of prehistoric Native Americans never proceeded beyond the stage of intermediate disturbance, their activities nevertheless changed the landscape mosaic, with lasting effects through at least one cycle of forest succession.

FOREST FRAGMENTATION AND PERCOLATION THEORY

Landscape ecologists use percolation theory to predict how close a landscape is to reaching critical ecological thresholds such as the connectivity of habitats (Gardner and O'Neill, 1991; With and Crist, 1995). Given a mapped landscape pattern represented as a two-dimensional grid in which grid cells are classified according to habitat type, percolation theory predicts that a random distribution of a single cell type that comprises at least 0.59 of the landscape has a very high probability of spanning the map (Gardner *et al.*, 1987). A high degree of connectedness between forest patches enhances gene flow between populations of terrestrial species, thus maintaining viable populations within self-perpetuating communities. Below this critical threshold, the habitat becomes isolated into numerous, discrete clusters. The landscape may become disconnected as the "backbone" of the percolating cluster is broken by removing critical habitat cells along the spine, separating the cluster into two separate habitat patches. This transition occurs abruptly and, on natural landscapes, may result in disjunct (allopatric) populations of animal and plant species. Even small changes in the composition of the landscape mosaic that occur near the critical threshold may have major effects on the distribution and persistence of plant and animal populations (Turner and Gardner, 1991; Turner *et al.*, 2001).

Mississippian-age agriculturalists and habitat continuity

By comparing landscapes from preagricultural time with those dating from the Mississippian cultural period – for example, in the Central Mississippi Alluvial Valley – we can estimate how close Mississippian-age agriculturalists were to exceeding the critical threshold of connectivity in the forest mosaic. In this way, we can evaluate the importance of prehistoric human impacts in potentially destabilizing the environment. In the Central Mississippi Alluvial Valley, *c.* 1000 BP, Mississippian settlements were extensive across most of the levee crest habitats in the active meander train (Morse and Morse, 1983). Forests that in Woodland times (*c.* 2000 BP) occupied those bottomland habitats were highly fragmented (percolation connectivity threshold $p = 0.2$). For deciduous forest and upland oak–hickory forest on braided-stream terraces of the Western Lowlands, however, $p = 0.8$ in Mississippian times, because human occupation was both less extensive there and was localized along tributaries to the Mississippi River such as the Cache River in eastern Arkansas (Delcourt *et al.*, 1999).

The vacant-quarter hypothesis

An important question in eastern North American archaeology concerns the causes for changes in settlement patterns associated with the development and decline of Mississippian cultures between about 1150 and 450 BP (Griffin, 1978, 1984, 1985; Williams, 1977, 1982). The "heartland" for development of Mississippian settlement systems (Figure 7.1) was located along the Central Mississippi Alluvial Valley from west-central Illinois (the Cahokia metropolis in the American Bottom; Milner, 1998) through southeastern Missouri (Powers Fort Site; Price, 1978, 1982) to east-central Arkansas (including the Parkin Site; Morse and Morse, 1983). To the east, the maximum development of Mississippian cultures extended into the lower Ohio River Valley, the Tennessee River Valley drainage as far east as eastern Tennessee (the Little Tennessee River Valley; Chapman, 1994), and southward along the Black Warrior River into central Alabama (Moundville; Peebles, 1978).

Mississippian people were sedentary and agriculturally based, dependent upon cultivation, first of maize and squash and then of beans, within alluvial bottomlands (Smith, 1978). At the height of Mississippian cultural development, typically between 900 and 500 BP, ceremonial centers with earthen temple mounds were constructed adjacent to extensive plaza areas. Numerous smaller villages with platformed mounds were dispersed over a wide region. The influence of Mississippian people extended throughout much of the eastern United States through development of extensive trade networks (Williams, 1982).

Large areas of the Central Mississippi Alluvial Valley were abandoned by Indian populations by 550 BP, leaving a "vacant quarter" (Williams, 1982) or "empty quarter" (Smith, 1986) within the former heartland of Mississippian settlements (Figure 7.1). With a population shift from braided-stream terraces to the meander belts of the Mississippi River, most of the remaining population of Native Americans became "nucleated," or concentrated, within fortified villages near prime agricultural land (Morse and Morse, 1983). By 400 BP, at the time of the first Spanish explorations of the region (Hudson, 1997), most of the Central Mississippi Alluvial Valley was controlled by six major chiefdoms that consolidated into a single tribe of "Quapaw" by the time of French explorations around 270 BP (Morse and Morse, 1983).

Several alternative hypotheses have been proposed to account for the phenomenon of the vacant quarter: (1) depletion of soil fertility with increasingly intensive cultivation of maize; (2) climatic changes associated with the Little Ice Age; and (3) warfare (Williams, 1982; Muller, 1982; Price, 1982; Smith, 1986). The degree to which climate change was responsible for abandonment of the vacant quarter is uncertain. Little Ice Age climate-cooling began after 600 BP (Lockwood, 1979) and thus largely postdated the time of decline in Mississippian cultures in the Central Mississippi Alluvial Valley.

In the Powers Fort area of southeastern Missouri, after about 580 BP, human impacts on native vegetation that resulted in deforestation, followed by unsuccessful attempts at cultivation, may have depleted organic carbon in the soil. Abandoned garden plots may have been invaded by sod-forming grasses that could have prohibited further attempts at cultivation using stone hoes. Difficulty in keeping fields open for cultivation may have contributed to the decision to abandon Mississippian sites (Price, 1982).

Case studies from two areas in the Mississippian heartland, the Little Tennessee River Valley in eastern Tennessee, and the American Bottom of southern Illinois, document in detail the record of human settlement, expansion of agricultural ecosystems, and changes in environments within the area of influence of Mississippian people.

THE LITTLE TENNESSEE RIVER VALLEY

The Little Tennessee River Valley provides one of the premier examples of prehistoric human–land interaction documented from North America. When, where, and how did Native Americans use landscapes along the lower stretch of the Little Tennessee River Valley? How did the passage of late-Quaternary thresholds on climate, geomorphologic processes, river regime, and vegetation shape the long-term pattern of human opportunities, adaptation to, and cultural transformation of natural ecosystems?

Physical setting and late-Quaternary history

The Little Tennessee River flows northwestward from its headwaters along the Eastern Continental Divide in western North Carolina, passing through steep V-shaped gorges cut through the igneous and metamorphic bedrock of the Southern Blue Ridge Mountains, then joining the broader Tennessee River within the Great Valley of eastern Tennessee (Figure 7.1). As the Little Tennessee River leaves the highlands and enters the valley, it migrates laterally, carving out its own wide alluvial valley from the gently rolling uplands and ridges of carbonate and shale bedrock. Nine sets of ancient river terraces stair-step down these hill slopes and flank the modern river channel. Sediments of the terraces were eroded from upstream highlands and were deposited during past glacial–interglacial cycles of the Quaternary.

In peak cold times of ice-age climate, for example from 27 000 to 17 000 BP, the summits of the nearby Southern Blue Ridge Mountains were periglacial settings, with alpine tundra and snowfields above 1450 m elevation, and boreal forest of spruce (*Picea*) and jack pine (*Pinus banksiana*) covering mid- and lower hill slopes down into the Great Valley (Delcourt and Delcourt, 1998a). With intensifying seasonal contrast, from 17 000 through 11 500 BP, permafrost melted and released earthen debris downslope into tributary streams. The Little Tennessee River filled in its active channels and the floodplain aggraded, rising by 7 m between 17 000 and

7800 BP. As boreal conifers were replaced by deciduous trees, development of closed forest diminished sediment erosion from hillslopes by 11 500 BP. With a marked reduction in sediment supply, the Little Tennessee River stabilized in a new equilibrium condition between 7800 and 4000 BP (Delcourt, 1980b).

Renewed channel incision after 4000 BP caused the abandonment of the former floodplain, dropping the groundwater table and stranding Terrace 1. In the late Holocene, the active stream channel became anchored within a narrow chute of seasonally inundated floodplain. Destabilization of the stream due to changes in its hydrology 4000 years ago may have been a combined result of human land clearance and a change in atmospheric circulation patterns (Delcourt and Delcourt, 1984; Mills and Delcourt, 1991).

The archaeological record

Prehistoric people left a direct record of their lifeways within the accumulating layer-cake of alluvium spanning the past 14 000 calendar years. For the late Pleistocene and most of the Holocene, the annual pattern of springtime flooding in the aggrading Little Tennessee River provided nutrient enrichment for productive bottomland forests. Native Americans used this riparian corridor for movement, exploitation of plant and wildlife food resources, and quarrying high-quality cherts from locally exposed rock outcrops to make stone tools.

As part of the archaeological salvage prior to completion of the Tellico Dam, the Tennessee Valley Authority (TVA) sponsored major excavations within the area to become impounded by the Tellico Reservoir, as well as on uplands up to 5 km from the Little Tennessee River. An extensive survey documented a total of 624 aboriginal sites that were used to identify single or multiple prehistoric occupations by specific cultural groups (Chapman, 1975, 1994). Of these sites, 337 contained diagnostic stone tools (65 422 artifacts) or ceramic pottery fragments (39 457 sherds). These data are compiled by the temporal sequence of archaeological "components" associated with specific time intervals for cultural phases (Kimball, 1985; Davis, 1990). Analysis of 21 cultural phases documented for the past 14 000 years thus gives insight into the panarchy of human ecosystems as they evolved in the Little Tennessee River Valley during the late Pleistocene and Holocene.

Chronology of human occupation

A total of 126 absolute dates documents human presence from archaeological sites: (1) 74 radiocarbon dates analyzed from organic matter (Kimball, 1985; Davis, 1990); and (2) 52 archaeomagnetic (AM) dates obtained from baked clays of prepared hearth bases and pottery sherds (Lengyel *et al.*, 1999). Additional radiocarbon dates establish the local geomorphic history and paleoecological records of vegetation change (Delcourt, 1980b; Delcourt *et al.*, 1986).

Similar trends are apparent for changes in intensity of settlement and archaeological site distribution through time (Figure 7.3). Initial exploration of the Little Tennessee River Valley occurred during the Paleoindian

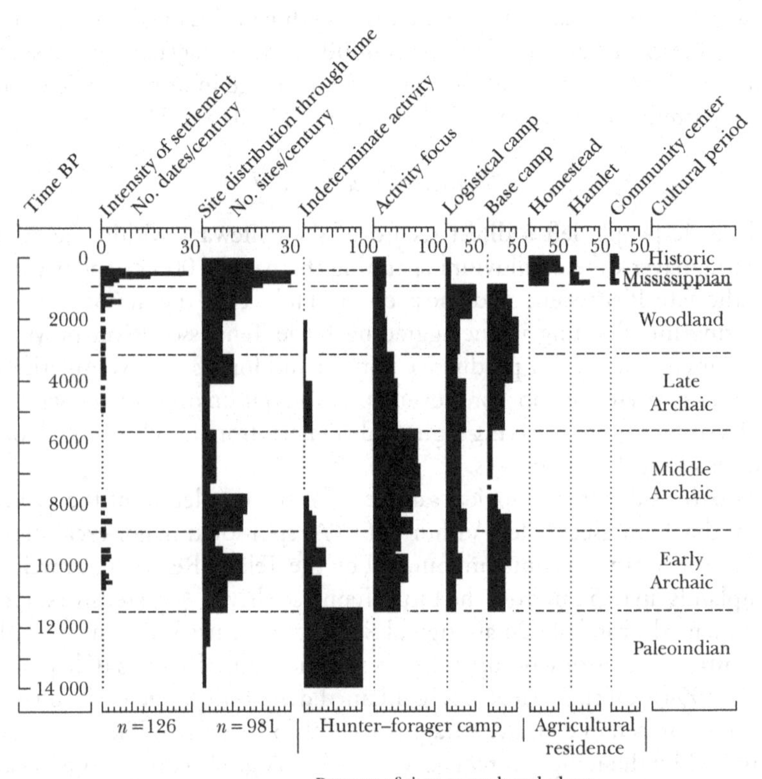

Figure 7.3 Late-Pleistocene and Holocene human populations within the Little Tennessee River Valley, eastern Tennessee. The 14 000-year chronology of settlement intensity is documented as numbers of absolute dates ($n = 126$) from archaeological sites plotted per century. A second proxy for prehistoric human populations is the number of identified archaeological site components ($n = 981$), standardized by duration of each cultural phase, and plotted per century. Tallied as percentages for each of 21 cultural phases, the types of occupation sites demonstrate the long-term shift from hunter–forager camps to agriculturally sustained residences (figure based on data from Davis, 1990).

interval, from 14 000 to 11 500 BP. Intensity of human occupation increased in the Early Archaic cultural period, from 11 500 to 8900 BP, then declined in the Middle Archaic, and rebounded after 4000 BP as populations grew and became sedentary. Numbers of both absolute dates and of components rose to the highest levels in the Woodland cultural period, after 3200 BP, and at the start of Mississippian lifeways, *c.* 1000 BP. The rapid fall-off in numbers within the past two centuries marks the forced removal of the Cherokee to their Oklahoma reservation 117 BP.

Material evidence of human activities

Temporal patterning of different kinds of archaeological sites is material evidence that reflects changes in human activities from Paleoindian to Historic times (Figure 7.3). Site use is interpreted on the basis of the numbers and kinds of artifacts found, the evidence for hearths and home structures, and numbers of features denoting intensive and recurring use. Overall, in the Little Tennessee River Valley, there was a long-term shift from primary transitory forays for gathering food to increasingly regular use of seasonal base camps in favored settings. Following these changes, year-round settlements supported by agriculture were established.

The hunter–gatherer mode of foraging is linked to exploratory movement across large territories, identifying valuable locations offering food, shelter, and stone quarries, and establishing a regular seasonal round of movement among the resource patches. Logistical foraging sites are categorized by several types of inferred activity: (1) indeterminant because of insufficient evidence; (2) an activity focus centered upon hunting, fishing or gathering of mast nuts and other edible plants; and (3) logistical camps for short-term encampments (Figure 7.3). Foraging base camps are identified on the basis of substantial or concentrated evidence for semi-permanent, although seasonal, residence. Permanent, year-round, agricultural residences include single-family homesteads, multiple-family hamlets, and local community centers with villages numbering up to 200 inhabitants (Davis, 1990).

The landscape location of logistical foraging sites and residential bases of operations varies in location with time across the environmental spectrum of habitats available from valley bottom and terraces of the Little Tennessee River, the Tellico River, upper-tributary watersheds, and the surrounding uplands. Two modes of landscape utilization are evident during the past 11 500 calendar years: (1) residential bases clustered along the active floodplain and lower terrace; and (2) dispersal forays into the hinterlands on seasonal quests for fresh game and nuts. Between 4000 and 600 BP, expanding human populations within the Little Tennessee River Valley resulted in the fissioning off of new satellite communities, established as outposts farther up the Tellico and other tributary bottoms. After 600 BP, outlying homesteads pulled back into town centers on low stream terraces, perhaps because people were attracted by arable land and because the perimeters of their communities could be bounded by stockades to offer more strategic defense in the event of warfare (Davis, 1990).

The prehistoric archaeological record of diagnostic weapons and tools can be interpreted in the context of how native people interacted with their local ecosystems within the Little Tennessee River Valley. Stemmed projectile points used for darts and bannerstone weights are evidence that the atlatl (spear thrower) was used for hunting from 14 000 to 1500 BP. Late Woodland and Mississippian people adopted the use of bows whose arrows were tipped by small, triangular stone points. Bow-hunting efficiency is tied to the length of line of sight. Bow hunting therefore might have become practical only after the landscape was opened by cumulative clearance past the percolation threshold of forest connectivity. Notched-pebble net sinkers and bone hooks indicate that people relied on fishing as early as 8900 BP. Fishing the river shoals was an especially important activity from 4200 to 3200 BP, the interval during which the broad, shallow channel of the aggrading floodplain gradually stabilized and then started to downcut (Chapman, 1994).

Wood-shaping tools, requiring attention to fine detail and accuracy, first occurred almost 11 500 calendar years ago. The abundance of heavy-duty, working-wood tools such as grooved axes, ground-stone celt axes, and chisels from sites dating after 10 700 BP indicate intentional tree-cutting. Horticultural tools such as chipped-stone hoes and wooden digging sticks date as early as 7800 BP in lowland sites and are commonly recovered from sediments postdating 2200 BP (J. Chapman, personal communication).

Ancient hearths were constructed as either basin-shaped pits filled with fire-cracked rock or were prepared over a clay pad. Evidence for deliberate processing of plant bast fibers is demonstrated as impressions of woven fabric, cordage, or basketry preserved on surfaces of hardened clay fragments. Although technology for creating handspun yarns and plaited fabrics existed as far back as nearly 11 500 years ago (Chapman and Adovasio, 1977), their impressions on clay surfaces became consistently and abundantly preserved only with the manufacture of ceramic pottery established locally within the last 3 200 years. The case for prehistoric processing of plant foods is inferred from durable tools such as grinding manos, milling slabs, pestles, and nutting stones, which were all items of heavy "site furniture" left at seasonal base camps.

An annual cycle of native plant harvesting and of seed parching for preservation and storage of plant food is also linked with the need for seed caches that were probably intended as safety food reserves and seed stocks saved for sowing crops in the next year's growing season. Increasing reliance upon stored plant food is consistent with the archaeological record of durable containers. Such containers included steatite-soapstone vessels with organic residues dated from 4000 to 3200 BP (Sassaman, 1999), and ceramic pots thereafter. After 2000 BP, archaeological sites included increased concentrations of storage pits that were often filled with caches of seeds.

From the late Pleistocene through the mid-Holocene, individual homes were temporary shelters that left little archaeological trace beyond hearths and the scatter of artifactual debris. More permanent structures are first documented in 3 200-year-old base camps (Figure 3.3). These structures are evidenced by post holes where building timbers were positioned, and by clay-daub fragments of wall coating. By 1000 BP, architectural "footprints" for homes contained rectangular post-hole patterns of wooden supports. By mid-Mississippian times, permanent dwellings were typically adjacent pairs of winter/summer homes.

Civic architecture of political chiefdoms involved substantial labor investment in moving earthen material. Irregular ridges were leveled for municipal plazas and ball fields. Starting about 1500 BP, Native Americans carried basket loads of sediment to build up conical and platform-shaped mounds. By 1000 BP, the centralized community organization created the need for exploiting more wood. Trees were cut and trimmed to produce circular townhouses and to bolster community defenses with palisade barriers (Chapman, 1994).

Trends in settlement patterns

Several fundamental trends of aboriginal settlement occurred within the Little Tennessee River Valley (Davis, 1990). From Paleoindian through Archaic times, until about 4000 BP, fluctuations in both site intensity and frequency reflect limited, ephemeral use of residential base camps, seasonally revisited at preferred locations along the valley. Logistical sites and activity foci document broad search patterns for garnering valuable resources from the upper tributaries and interfluves. These initial colonists had an annual round of movements through a territorial range extending well beyond the Little Tennessee River. Although ecologically tethered to this prominent riparian corridor, small bands of people moved widely among the major physiographic provinces of the Great Valley, the Southern Blue Ridge Mountains, and possibly as far as the Carolina Piedmont.

As residential population sizes grew between 4000 and 1000 BP, the overall territory of the band decreased and became focused on the landscape ecotonal transect from the Great Valley to the Southern Blue Ridge Mountains. Late Archaic and Woodland-age sites show more intensive use of local resources, involving fishing, forest fragmentation to establish cleared areas, and horticultural experimentation. The seasonal round of foraging shifted away from the former river route and radiated from the bottomlands into the upland zone running parallel to the Little Tennessee River.

Within the past 1000 years, rapid population growth mirrored the emergence of maize agriculture as the primary subsistence base, as well as the development of societal distinctions among homestead, hamlet and local community centers. With the social organization of a ruling chiefdom, year-round residential settlements relocated from the lowest stream terrace to the next higher alluvial bench. This shift in Mississippian residential location has been interpreted in the context of the need for larger and sustainable quantities of food, and of imminent ecological threat. Population increases accelerated the need for productive, arable land. As the alluvial land became more valuable for growing maize, people moved their towns up slope. Population pressure fueled the spread of extensive croplands as well as the intensifying exploitation of wood supplies needed for hearths, homes, and constructing defensive palisade walls around town centers. Denuded hill slopes and alluvial bottoms accentuated overland flow of precipitation runoff, amplifying flood frequency and severity. For example, dated from the interval 850 to 400 BP, sediment layers burying archaeological sites provide tangible evidence that residences on the lowest stream terrace experienced major flooding episodes (Schroedl *et al.*, 1985). This Mississippian lifeway terminated with the historic European American treaties and forced removal of the Cherokee Nation.

Paleoecological evidence for human impact

Paleoecological evidence for human impact has been amassed from extensive paleoethnobotanical studies, and pollen, charcoal, and plant macrofossil analysis of sediments from small ponds located both adjacent to known archaeological sites in the Little Tennessee River Valley as well as in the uplands distant from known centers of prehistoric agriculture.

These data are supplemented by analysis of faunal remains in archaeological contexts.

Ethnobotanical assemblages of plant remains excavated from archaeological sites (Chapman *et al.*, 1982) and fossil–pollen records of vegetation change preserved in sediment cores from small ponds (Delcourt *et al.*, 1986) show the consequences of human use of plant resources over the past 11 500 calendar years (Figure 7.4). Wood–charcoal spectra from hearths include 37 types of trees in addition to woody culms of American cane.

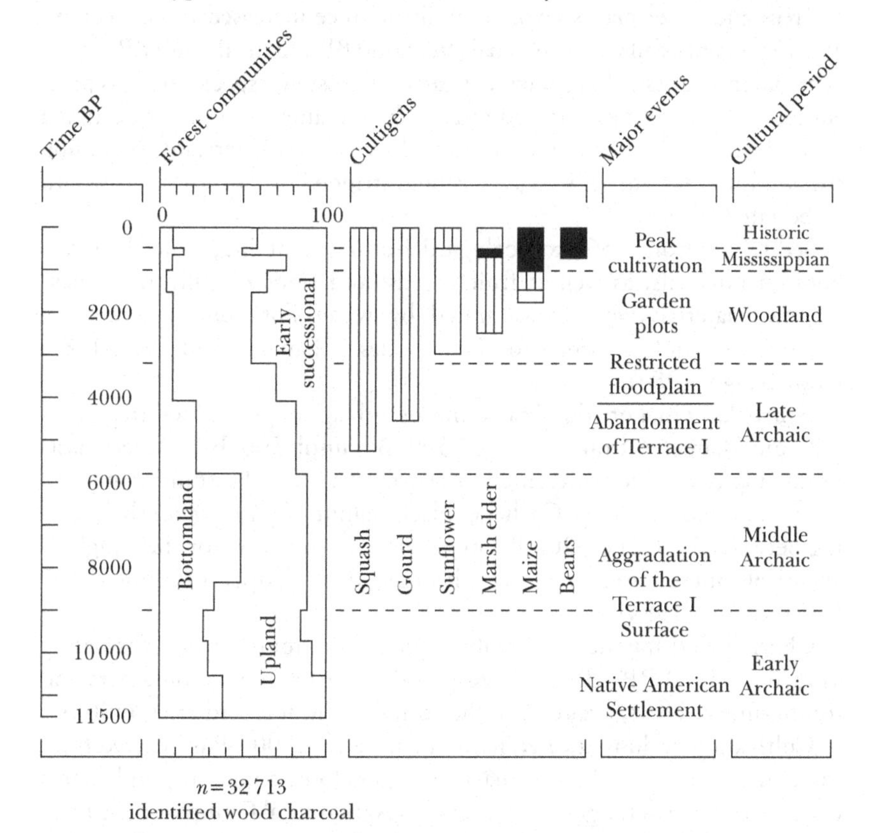

Figure 7.4 Evolution of prehistoric human ecosystems of the Little Tennessee River Valley, eastern Tennessee, represented by changes in Holocene fluvial regimes, compositional change in forest communities, and the sequence of Native American cultivation of plants. Forest communities are characterized from assemblages of wood charcoal ($n = 32\,713$ fragments of wood charcoal identified to species). For plants represented in ethnobotanical samples, the open vertical bars indicate rare occurrence, vertical stripes indicate common occurrence, black bars denote abundant occurrence (modified from Chapman *et al.*, 1982).

The macroscopic wood–charcoal remains represent the types of materials collected for fuelwood in the vicinity of base camps or villages.

From 11 500 to 4000 BP, the fuelwood was an equal mix of bottomland and upland species. The charcoal assemblages contain only 8 percent floodplain species after 4000 BP, however, and less than 5 percent after 1500 BP. The decline in representation of bottomland species probably reflects both the progressive denudation of the lower stream terraces and stream entrenchment that would have shrunk the available floodplain habitat.

Trees and other plants favored by disturbance increased to between 30 and 45 percent of all wood charcoal after 4000 BP and until 1500 BP. Wood charcoal in this assemblage includes early-successional species such as pine, American cane, and eastern red cedar. These plants would have occupied areas cleared, cultivated, and abandoned by Native Americans, including sites on lower terraces as well as hillslopes stripped by soil erosion down to carbonate bedrock.

Sediments from 956 archaeological features contained abundant carbonized nutshells, as well as fruits, seeds, and rinds of cultigens. These organic materials were concentrated by water flotation, separated by size fraction, and identified to plant species (Chapman and Shea, 1981; Cridlebaugh, 1984).

Nutshells of hickory dominated the assemblage of plant–food fragments >2 mm diameter from nearly 11 500 BP until 500 BP. Other mast-producing trees were represented by acorns, walnuts, chestnuts, American beechnuts, and hazelnuts. Caches of black walnut (*Juglans nigra*) shells dating between 5200 and 3200 BP might have been stored for their juglone chemical content, used to poison fish stranded in floodplain "fishing holes" (Johannessen, 1984).

Charred rind fragments of edible squash were found in hearths dating as early as 5200 BP. Hard-shelled gourd, used for bottle containers and fishing-line bobbers, occurred in the ethnobotanical record after 4500 BP.

Cultivation of domesticated native plants after 2700 BP is inferred from the abundance of sunflower, maygrass, goosefoot, amaranth, and marsh elder. Use of these cultigens of the Eastern Agricultural Complex continued well into the Mississippian cultural period, with marsh elder and knotweed (*Polygonum*) comprising up to 60 percent and 30 percent, respectively, of the small-seed assemblage.

Although the oldest specimens of maize date from 1700 BP (Chapman and Crites, 1987), use of corn in the diet was minor in the Little Tennessee River Valley during the Woodland cultural period. Only after 1000 BP did the proportion of maize rise relative to mast nuts. Maize constituted as much as 50 percent of the ethnobotanical remains by the Late Mississippian cultural period, reflecting an increased dependency upon maize in the diet and the decreased use of native domesticates and nuts. Beans were introduced into the lower Little Tennessee River Valley about 600 BP. Soon thereafter, Spanish conquistadors introduced the peach (*Prunus persica*), which was grown widely in historic Cherokee orchards.

The paleoecological record of pollen grains and plant macrofossils has been studied from late-Holocene sediments of small pools in the Little Tennessee River Valley (Delcourt *et al.*, 1986). Tuskegee Pond, situated on the third terrace above and 1.5 km distant from the active floodplain of the Little Tennessee River, contains organic-rich sediments spanning the past 1500 calendar years. The record of 1 to 2 percent maize pollen preserved throughout the sediment sequence demonstrates that maize was cultivated in fields adjacent to the pond throughout the Late Woodland and Mississippian cultural periods. Generally, higher maize values occurred after 1000 BP. Goosefoot and marsh elder, cultivated during Woodland times, were also represented in the fossil pollen record. Accumulation rates of charcoal particles and influx of mineral sediment into Tuskegee Pond both increased by an order of magnitude at about 1000 BP, corresponding with the transition from Woodland to Mississippian cultural periods. We interpret the major increase in charcoal accumulation rates to reflect increased use of fire for swidden agriculture. Conversion of forest to agricultural fields would also have increased soil erosion considerably above levels imposed previously by less intensive farming practices (Delcourt and Delcourt, 1998a).

Ragweed is a ruderal plant that invades agricultural fields and provides a measure of the area of land in cultivation. Pollen percentages of ragweed are generally <5 percent of the pollen sum in modern samples from forested landscapes of the eastern United States, but they reach much higher values in surface samples from landscapes in the Midwestern United States that have undergone widespread land clearance for agriculture (McAndrews, 1988). Fossil pollen percentages for ragweed therefore serve as a proxy for the extent of anthropogenic forest clearance in the past. At Tuskegee Pond, ragweed pollen was 20 percent of the pollen sum at 1500 BP, and it increased to over 50 percent after 600 BP. Taken together, the fossil pollen records of maize, goosefoot, marsh elder, and ragweed support the interpretation that Late Woodland and Mississippian people maintained extensive agricultural fields on stream terraces extending as far as 1.5 km from the active floodplain of the Little Tennessee River.

Assemblages of tree pollen indicate changing species composition and structure within progressively fragmented forests near Tuskegee Pond.

Between 1500 and 1000 BP, pollen of bottomland trees declined, reflecting preferential clearance of lower stream terraces for garden plots. Pine pollen increased steadily in the Tuskegee Pond record from 25 percent of the arboreal pollen at 1500 BP to a peak value of 45 percent by 850 BP. Pine pollen declined to about 20 percent in Late Mississippian times. We interpret rising values of pine pollen between 1500 and 850 BP to reflect the invasion of southern pines into the patchwork of Native American old-fields scattered along the Little Tennessee River Valley. After 850 BP, maize remained consistently represented by pollen, and ragweed percentages remained high. Percentages of pine pollen, however, declined. We interpret the decline in pine pollen during the Mississippian cultural period as resulting from an expansion in the area of permanent village sites. Mississippian people evidently maintained permanent croplands but cut adjacent pine stands, both to enlarge their agricultural fields and to harvest softwoods for pole timbers that were incorporated into new homes and palisades.

Even in the increasingly deforested alluvial valley, upland knolls apparently harbored remnant stands of hardwoods. Mississippian-age assemblages of tree pollen remained dominated by oak, hickory, and chestnut. Substantial food-storage caches contained abundant ethnobotanical remains of mast nuts. We interpret these data as evidence that upland orchards of oak, hickory, and chestnut trees were considered by Native Americans to be "sacred groves." Mississippian people continued to rely on mast nutmeats to complement their maize crop. Declines in pollen percentages for these hardwood species date from 190 BP and the construction of British Fort Loudoun, located 1 km north of Tuskegee Pond. The Colonial English preferred oak and hickory timbers for building their fort palisades, and they stripped the last mature forests on the upland bedrock knoll directly west of the pond.

The case study from the Little Tennessee River Valley documents the evolution of a self-organizing, panarchical human ecosystem through the Holocene interglacial. After aboriginal Native Americans began to cultivate native weedy plants along the Little Tennessee River some 5200 years ago, they began to rely upon crop yields that were sustained and predictable in quantity, nutritional value, and storability. As human populations grew, they used wildfire to clear more and larger forest gaps. As garden plots were used and abandoned, they were invaded by opportunistic pioneer plants such as pine, eastern red cedar and cane. Conversion of once-contiguous forested tracts to a fine grained mosaic of disturbance-generated patches increased the extent of ecotonal edge habitats in which mammals such as white-tailed deer would have increased their populations. With increasing human impact, larger quantities of plant food could be grown, and thus wildlife populations initially would have been favored.

With increasing forest fragmentation, however, by Late Woodland and Mississippian times, the threshold for habitat continuity of native bottomland forest communities was exceeded. With ever-increasing Mississippian human populations, even relatively young pine stands were cut, burned, and converted to farmland. Nearly all of the area on the terraces adjacent to the Little Tennessee River was cleared for agriculture, and the annual replanting of crops was a repeated and intensive disturbance regime that allowed for only minimal secondary succession of cane, weedy herbs, and shrubs. Thus, in Mississippian times, the landscape as a whole was in a disequilibrium state caused by anthropogenic activities. Human needs for animal protein also intensified. In the archaeological record from the Toqua site, dating *c.* 650 to 150 BP, faunal remains of white-tailed deer document increasing hunting pressure as hunters preferentially took the largest available bucks, and the white-tailed deer populations were over-harvested to near extinction (Davenport, 1999).

A dynamic model of Mississippian agricultural ecosystems

Using a dynamic simulation model, Baden (1987) tested the ecological implications of Mississippian-age settlements within the Little Tennessee River Valley. Based upon archaeological excavations dated *c.* 1050 to 250 BP, the full Mississippian cultural sequence for prehistoric maize production was simulated as an interactive model driven by expanding numbers of resident aboriginal populations, their cumulative requirements for minimum food quantities necessary to feed them, and the increasing demand for more suitable areas to be placed under cultivation (Figure 7.5). This modeling approach applied stability theory, assessing vulnerability of prehistoric maize agricultural ecosystems to plausible and archaeologically defined scenarios of climatic variability, ecological limits, and behavioral constraints tied to technological and societal dependence in maintaining a maize-based subsistence (Baden, 1987).

The system-based model presumed a climatic regime with at least a 120-day frost-free period necessary for primitive races of maize to ripen. The primary ecological limit for cultivation was set at the maximum area of soils suitable for growing crops using primitive stone hoes. Demographic boundary conditions assumed suitable seed stocks were available, that prospective farmers already had acquired the necessary horticultural knowledge, and

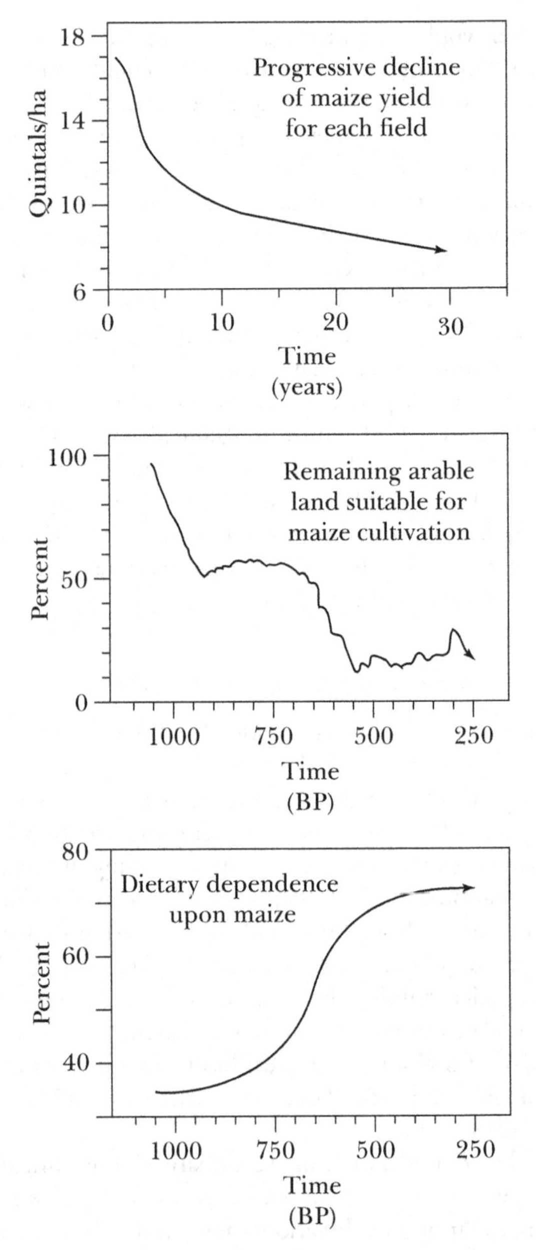

Figure 7.5 Mississippian model of expanding prehistoric agricultural ecosystems, environmental degradation, and dietary needs for Native American populations in the Little Tennessee River Valley, eastern Tennessee (modified from Baden, 1987).

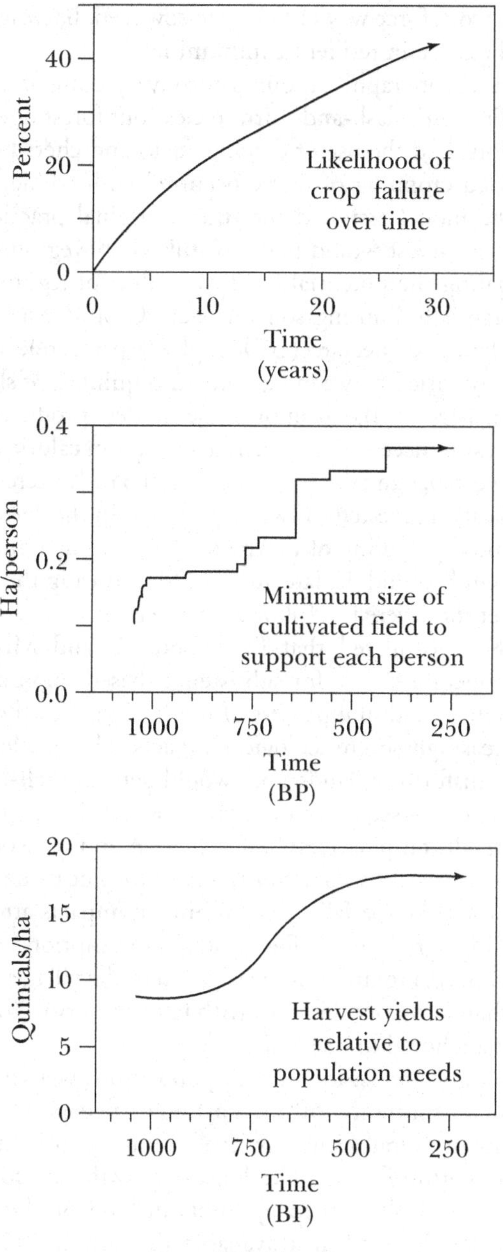

Figure 7.5 (Cont.).

that the size of work force was sufficient to sow seed, harvest, and store the crops required to sustain resident communities.

Early historic ethnographic accounts portray a picture of traditional aboriginal use of fire, with slash-and-burn to clear out forest openings for new agricultural plots, and the use of digging sticks and chert hoes to prepare seed beds. Mixed cropping of maize occurred with edible, fleshy squash and beans. The model assumed that the aboriginal practice of burning these fields after harvest would have volatilized aboveground phytomass, effectively negating the potential fertilizing effect of legume beans in replenishing potentially limiting soil nitrogen. Crop densities approached 4050 maize plants per hectare (10 000 plants per acre), and aboriginal harvests typically varied between 6.3 and 12.6 quintals of shelled corn/ha (10 to 20 bushels/acre). Based upon these modest production levels, the minimum field sizes necessary to generate minimum calorie needs for food were set at between 0.1 and 0.6 ha per person (0.3 to 1.5 acres/person). The maize was typically harvested in two stages: first, in the "green" corn stage for a mid-summer gathering of unripe ears, providing a guaranteed minimum supply; and second, in late autumn for ensuring the stored reserve needed to offset the severe food shortages of winter.

Baden (1987) postulated that for Woodland and Mississippian-age people, the deliberate choice for subsistence, based upon an agricultural lifestyle, generated a cultural process of interaction with the environment that led to increasingly severe ecological impacts. This accelerating cultural trajectory in transforming landscapes would generate self-imposed instabilities, as a direct consequence of environmental degradation caused by traditional agricultural practices. Soil depletion and firewood exhaustion provided the incentive for village communities to relocate periodically, thus shifting in location the spatially focused human impacts attendant on this mode of "shifting agriculture." The general presumption was that Native Americans practiced a rotation in the location of the village and its planted fields, rather than a rotation in crops with fallow intervals for soil-nutrient replenishment anchored in one site.

The dynamics of prehistoric human populations were modeled with a high rate of infant mortality (50 percent), low life expectancy of 20 years from birth, and an overall fecundity level of four to five children for each reproductively mature female. The logistic growth rate for this population could have been dampened by limits in food production owing to constraints in (1) the land area available for farming; (2) the amount of suitable arable soils; (3) the available labor force; and (4) the amount of fertilizer added to replenish soil nutrients. Human populations would have been threatened by shortfall or failure of the maize crop. In turn, food shortages would have triggered behavioral responses such as wholesale relocation of villages, or reduction of food consumption in established communities by dispersal of families to establish new hamlets (fissioning). Practices of infanticide or territorial aggression could also have dampened intrinsic rates of population growth. Thus, expansion of agricultural fields would have generated a broader food base but would also have reinforced a feedback mechanism that could have magnified environmental impacts imposed initially by the establishment of an agricultural ecosystem.

Estimates of prehistoric crop-yield potentials were based upon early historic production of Northern Flint, and flour races of maize grown on the best alluvial soils, under optimal climatic conditions, using traditional aboriginal practices. Upper bounds for prehistoric yield of 18.8 quintals/ha (30 bushels/acre) were diminished by three key factors in the simulation model: (1) the local mix of soils and their fertility as mapped by modern soil surveys for the three counties comprising the area around the Little Tennessee River; (2) an exponential-decay function for depletion in soil fertility and fall-off in maize productivity, following swidden forest clearance and first-field planting; and (3) expectation of increased probability of crop failure with time and typical conditions of climatic variability associated with the Little Ice Age (Figure 7.5).

With a prominent decline in maize yield after about ten years of annual planting, it was presumed that the old-fields were abandoned for 100 to 150 years before soil fertility returned to preagricultural levels. This shifting patchwork of cultivation would lead to a long-term reduction in remaining arable lands available for growing maize. Model results indicated an initial 50 percent conversion of forested settings to farmland – that is, active fields and fallow old-fields – within one century of the onset of extensive and intensive, maize cultivation. The model simulated that between 950 and 650 BP, only half the suitable terrain was maintained in cultivation. The shifting patchwork of active garden plots and subsequent plant succession changed the landscape into a heterogeneous mosaic favoring disturbance-tolerant species. With the simulated expansion of active farming after 650 BP, more than 80 percent of appropriate soils might have been cropped annually, with little or no time for plots to have been set aside for replenishing soil nutrients. Rather, from 650 BP to the end of the simulation run at 250 BP, all of the former old-fields, even those with early-successional thickets and pine stands, were effectively cleared without allowing fallow "rejuvenation."

The minimum area of annually cultivated land necessary to feed each person nearly quadrupled over the 800-year span simulated for Late Woodland, Mississippian, and Historic time (Figure 7.5). The cultural solution to this problem required expanding the typical field size in order to satisfy increasing food needs that preferentially relied upon corn production. By the Late Mississippian cultural period, the dynamic model projected virtually contiguous agricultural fields across the entire extent of the floodplain and terraces of the Little Tennessee River. Long-term depletion of soil nutrients, however, imposed a steady drop in landscape carrying capacity, with progressively declining maize yields.

Stable carbon analyses of human skeletal remains are used as evidence that pehistoric diets became increasingly dependent upon maize (Lynott *et al.*, 1986). The dynamic model of Baden (1987) incorporated a demographic shift in the average quantities of maize consumed each year by individuals. This growing dependence upon maize food, expressed as a percentage of the total diet, exhibits a sigmoidal form, starting at 35 percent at 1050 BP, increasing to 50 percent by 650 BP, and reaching an asymptote of 70 percent after 450 BP (Figure 7.5). Using small field sizes (0.1 to 0.2 ha for each person), at first Mississippian farmers could have produced enough plant food to satisfy the community needs. However, with time, the annual harvest of shelled corn would not have been sufficient. Food shortfalls could have been triggered by insect damage and seed spoilage, frost or flood damage to growing plants, or raiding of stored maize reserves during episodes of warfare. Under these circumstances, even stable human populations (simulated as zero-population growth) would have been forced to change drastically in response to past experience and current perceived needs.

Without resorting to cultivating much larger fields, the trajectory of Mississippian populations would have encountered two critical points, potential omega phases, making the cultural system vulnerable to panarchical collapse (Figure 3.1). The first cultural threshold, at 950 BP, would have been reached when the population's need for land exceeded the finite amount of arable land within a several-kilometer radius of residential sites. The response to locally falling maize yields at that time could have been to fission off new satellite hamlets dispersed across the valley. The second cultural threshold would have occurred at 550 BP after large chiefdoms were established and when, effectively, all available, suitable soils were placed under prehistoric cultivation.

Based upon archaeological evidence, Baden (1987) projected that 200 individuals lived in each small village, and up to 1000 people lived in major Mississippian centers. Using a logistic model of population growth, Baden used a conservative intrinsic rate of growth of r = 0.003 and an initial prehistoric population of 200 within the Little Tennessee River Valley at 1050 BP. With 800 years of unchecked population expansion, a population level of 2200 would be expected by the early-Historic period at 250 BP. Using a census of Indian warriors and a map of Cherokee villages in the Little Tennessee River Valley, produced by British Lieutenant Timberlake at 193 BP, 2400 Native Americans were estimated to have lived in the lower valley of the Little Tennessee River in early-historic times. These early-historic population levels represented an ancient culture susceptible to crop failure and to territorial conflict, arising from their need to store or raid vital maize reserves.

The cultural instabilities in the Little Tennessee River Valley outlined by Baden (1987) were predicated upon an estimate of initial maize production of 17.2 quintals of shelled corn produced on each hectare of farmland (27.4 bushels/acre). That estimate was made on the basis of twentieth-century corn production in eastern Tennessee. As an alternative, Schroeder (1999) compiled data from the census of maize production by early-Historic Native Americans across the Eastern Woodlands and Great Plains, made by Schoolcraft in 113 BP. According to that census, maize yields would have been only about 6.3 quintals/ha (10 bushels/acre), reflecting the limitations of stone hoe technology and mixed-crop planting, as well as losses incurred because of rotting and seed predation. Such low maize production rates would have accentuated the aboriginal population pressures to transform landscapes into culturally maintained agricultural ecosystems and may have quickened the rate at which panarchical thresholds were approached.

CAHOKIA AND THE AMERICAN BOTTOM

The largest example of Mississippian settlement in eastern North America is the Cahokia metropolis complex, located in the American Bottom of southern Illinois (Figure 7.1) (Fowler and Hall, 1975; Bareis and Porter, 1984; Mehrer and Collins, 1995). The height of cultural development in the American Bottom was during the Stirling phase, from 800 to 700 BP (Milner, 1986; 1998), with maximum prehistoric populations estimated at 25 000 permanent residents in a 300 km^2 area (Milner, 1998). Human populations in the bottomlands near Cahokia underwent a progressive increase from the Emergent Mississippian period, 1100 to 900 BP, to a maximum during the Stirling phase, followed by a decline through the Moorehead and Sand Prairie phases, from 700 to 525 BP. By 500 BP, Mississippian people had abandoned the American Bottom, with only limited and temporary subsequent occupation by Oneota people from 500 to 300 BP (Milner, 1986, 1998).

The development of the Mississippian tradition in the American Bottom was an outgrowth of the previous 700 years of Late Woodland and Emergent Mississippian cultural change, rather than a dramatic revolution or replacement of local populations with people from outlying areas, (Mehrer and Collins, 1995). Permanent villages were first established during the Patrick phase at the end of the Late Woodland period, 1300 to 1100 BP. In the Emergent Mississippian period, 1100 to 900 BP, villages increased in community size, density, and complexity, with each family unit in a village using three or four functionally distinct buildings oriented around courtyards. Through the Lohman phase of the Mississippian period, 900 to 800 BP, rapid population growth was reflected by the appearance of town-and-mound centers, with residential districts oriented toward the cardinal directions and with communal granaries located in their plazas. As people moved from villages to mound centers, the settlement system evolved into a hierarchical settlement pattern, with large and small sites, monumental temple-mound architecture, and small isolated farmsteads in place of small villages. As the regional settlement hierarchy reached its peak of complexity, there was a change in the ways in which households related to each other and to their position on the landscape.

The platform mounds of the Cahokia complex were completed during the Stirling phase. The main Cahokia complex consisted of more than 100 earthen mounds distributed over 10 km^2 in the widest part of the Mississippi River floodplain in the American Bottom (Mehrer and Collins, 1995). Milner (1998) characterized Cahokia as the most important Mississippian-age archaeological site in eastern North America, based upon (1) the extent of habitation debris; (2) the intensity of occupation; (3) the presence of burials of highly ranked people; and (4) the monumental architecture. For example, Monk's Mound was a large platform mound which was oriented perpendicular to Cahokia Creek and formed the focal point of a central ceremonial precinct composed of a series of lesser platform mounds, all of which were surrounded by an extensive stockade wall.

During the Stirling phase, five additional, smaller multi-mound centers were also established, along with five single-mound centers, and a number of rural households were located within dispersed communities that were relatively isolated on the landscape. Archaeological evidence indicates that houses were rebuilt over original locations, reflecting long-term continuity of family lineages. Houses were characterized by central hearths and large, bell-shaped interior storage pits containing food and gardening tools, including chert hoes and white-tailed deer-antler rakes.

In the late Mississippian Moorehead and Sand Prairie phases, regional population declined, with activity concentrated on the highest-elevation sites in the floodplain, such as near Monk's Mound. Topographic setting became more important than social organization in the patterning of communities.

Models of Mississippian social organization

Two alternative models have been proposed for social organization of Cahokia. The classic model formulated by Fowler and O'Brien postulated a four-tiered hierarchy composed of three levels of functionally differentiated, platform-mound centers, with sites lacking mounds at the base of the hierarchy. This model conforms to the central place theory in which a centralized government controlled the manufacture and transport of trade goods, including projectile points and shell beads, as well as the import of food grown in intensive agricultural fields on the periphery of the city center (Butzer, 1982).

The alternative, contemporary model of settlement patterns (Milner, 1998) proposes a regional, decentralized settlement system, composed of structurally similar, potentially autonomous mound centers with nearby small village sites. Outlying mound centers would have been linked to Cahokia through political and social ties among chiefs who presided over self-sufficient territories centered on locally important mound groups. Outlying communities were composed of dispersed houses centered on special-function feature complexes. Thus, the floodplain of the Mississippi River in the immediate vicinity of the central Cahokia precinct would have supported a series of separate communities. Cahokia was a complex chiefdom, a society with minimal economic specialization, ranked and kin-based social groups, a limited number of inherited leadership positions, and chiefs who possessed great prestige and authority but probably had little true power (Milner, 1998).

Environmental factors constraining Mississippian cultural development

The meander train of the Mississippi River in the American Bottom was in general a poorly drained swampland. At the time of the General Land Office Surveys 100 BP, most of the floodplain was vegetated with dense

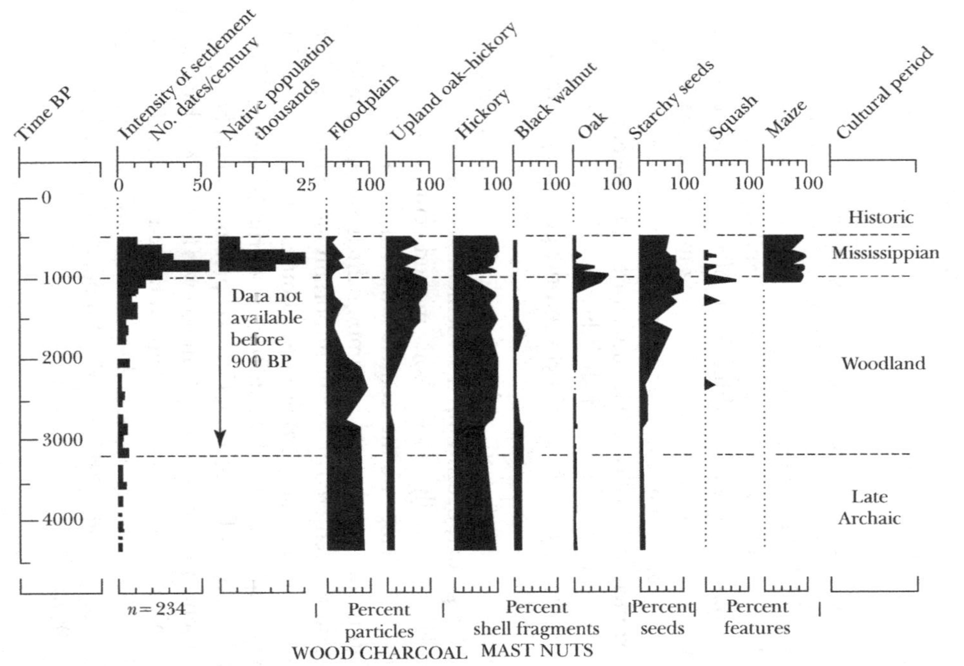

Figure 7.6 Prehistoric human ecosystems of the American Bottom, southern Illinois. Chronology of settlement intensity is expressed as numbers of absolute dates ($n = 234$) on archaeological sites per century and by estimates of population size (Milner, 1998). Evidence for forest clearance on floodplain sites is from wood charcoal. Ethnobotanical remains of nuts and seeds indicate activities such as foraging for mast nuts and cultivation of plants (data from Bareis and Porter, 1984; Fowler, 1989; Milner, 1998; Johannessen, 1984; Rindos and Johannessen, 1991).

floodplain forest, with wet prairie occupying limited areas of relatively high land, for example near Monks Mound (Milner, 1998). Local relief was 50 to 100 m from valley bottom to the top of bluffs bordering the floodplain about 4.5 km east of Cahokia. Loess-capped uplands supported both prairie and hardwood forest vegetation.

In the American Bottom, the positioning of settlements was strongly influenced by the landscape structure, including such factors as the geomorphic and hydrologic regimes as well as the vegetation. The most suitable sites for Mississippian people to live in the American Bottom included available expanses of relatively dry and potentially arable land, as well as old channel banks that were relatively high and close to the margins of oxbow lakes that could be easily traversed by canoe. These sites were densely occupied, and settlements also were located in areas of low ground that was regularly but temporarily flooded, adjacent to seasonally dry fish ponds. According to Milner (1998, p. 118),

> Suitable sites for houses in both mound centers and outlying settlements were tightly constrained by the disposition of dry ground. Ridges within the inner bends of old river meanders and natural levees were particularly good places to settle. There was no shortage of space for fields close to the late prehistoric sites, although fields near more heavily populated settlements, particularly Cahokia, must have been used more intensively than they were elsewhere.

Cahokia was thus strategically located in the middle of the widest part of the American Bottom floodplain, adjacent to one of the most extensive patches of relatively dry, arable soil, and close to Cahokia Creek and open-water oxbow lakes of the Mississippi River. This location had both a high proportion of suitable area for living and for cultivation of maize, and a high habitat heterogeneity, altogether giving the landscape a high capacity to support a locally large human population.

Archaeological record of population growth

The chronology of human occupation within the American Bottom (Figure 7.6) is documented by radiocarbon dates obtained on organic material sampled from features preserved within archaeological sites (234 dates covering the past 4500 calendar years, summarized in Bareis and Porter, 1984 [Appendix B]; Fowler, 1989; Milner, 1998). The great majority of these absolute dates was analyzed in conjunction with prehistoric sites excavated along the right-of-way on Interstate 270 in East St. Louis, prior to the construction of the 34-km stretch. Undated artifacts collected along the

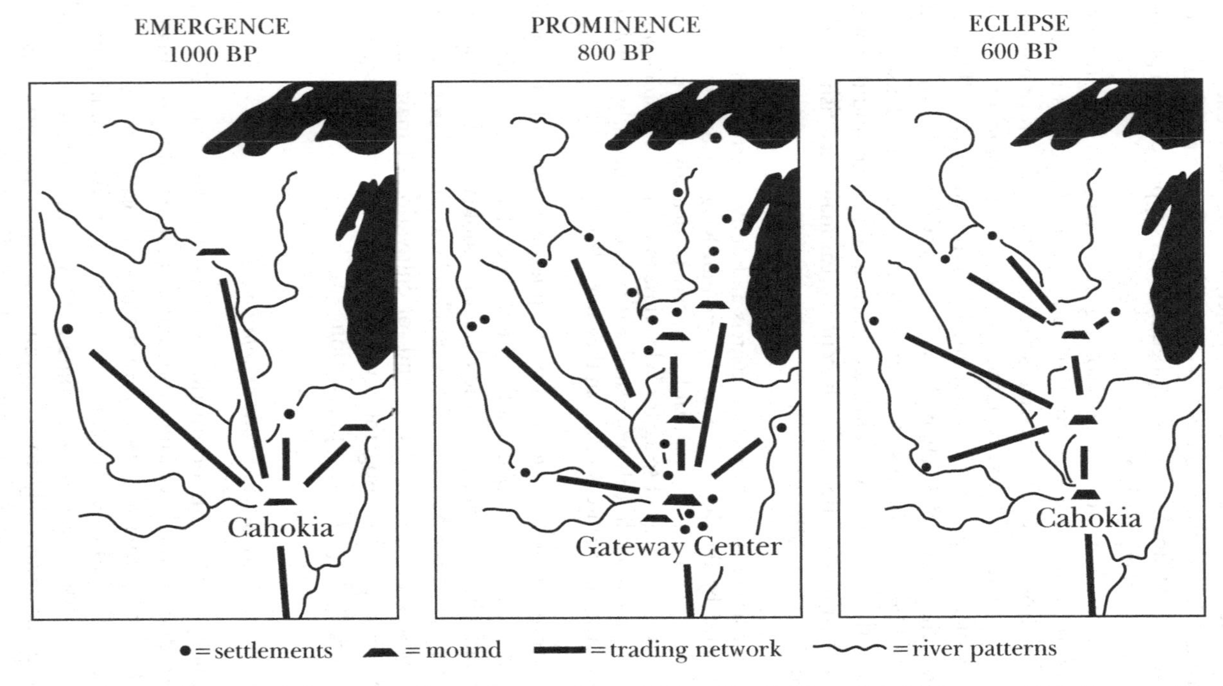

Figure 7.7 Prehistoric Cahokia metropolis as a Gateway Center 1000 to 600 BP. The political influence of Cahokia expanded as trading networks proliferated along transportation corridors of the upper Mississippi, Missouri, and Illinois rivers. Cahokia served as a market and trading center for the northern hinterlands until it was displaced economically by upstream market outposts (modified from Kelly, 1991).

bluff margin and nearby uplands document prehistoric human activities in Dalton and Early Archaic times (Esarey and Pauketat, 1992).

Within the American Bottom, the oldest absolute date from an archaeological context documents first human presence by about 6000 BP. After 4500 BP, the few available absolute dates (Figure 7.6) on archaeological remains reflect low human populations. An increase in number of radiocarbon dates from 3200 to 1500 BP probably reflects some combination of more continuous occupation and higher population with the first permanent settlement of the American Bottom. The abrupt rise to between seven and 15 absolute dates per century from 1500 to 1100 BP records the local expansion in settlements of Woodland-age people. The heightened variability in number of dates may be the result of incomplete archaeological census of prehistoric hamlet and village communities that were periodically shifted to new locales. A prominent increase in absolute dates between 1100 and 800 BP (Figure 7.6) parallels an exponential increase in Native American populations during the founding of Cahokia. The subsequent decrease in radiocarbon dates tracks the abrupt population crash after 525 BP (Young and Fowler, 2000).

The Emergent Mississippian period represents a time of population migration to a chain-like series of local centers that were typically associated with mound construction and situated on friable, sandy soils (Kelly *et al.*, 1984). Population growth at Cahokia reflected the centrifugal attraction for people moving from the uplands to agriculturally productive sites in the bottoms (Woods and Holley, 1991). This interval of population growth and establishment of towns (settlement nucleation) was linked with the onset of the widespread introduction and adoption of maize agricultural ecosystems and the emerging role of the Cahokia metropolis as a Gateway center (Kelly, 1991).

Cahokia played an important role in facilitating and regulating regional trade of (1) desirable raw materials such as salt, chert, hematite, galena, copper, catlinite, and steatite; (2) easily transportable foods such as maize; and (3) commodities fashioned by cottage industries of artisans, including plaited fabric (Drooker, 1992) and ceramic pottery. As both a market and urban center, Cahokia provided a central focus for trade networks extending along the transportation corridors of the Mississippi, Missouri, and Illinois Rivers (Figure 7.7).

With the expansion of Mississippian political control, establishment of northern market exchanges eventually compromised the prominence of Cahokia as a Gateway after about 700 BP. Onset of social disruption took place with the centripetal outmigration of the "Cahokia diaspora"

(Woods and Holley, 1991). Territorial conflict made construction of protective stockade walls around city centers necessary, and settlements began to re-establish across the uplands, relegating Cahokia primarily to the status of a ceremonial center.

Population estimates have been made for the time interval 900 to 525 BP for both the 10 km^2 core of Cahokia and the broader landscape setting of the 300 km^2 area of American Bottom, using several conservative criteria (Gregg, 1975; Milner, 1986, 1998). These criteria included (1) the tabulation of density of small, single-family homes for four different Mississippian-age cultural phases, each of which lasted from 50 to 125 years; (2) the presumption that the length of house use before decay of wall-post supports required replacement was no more than 5 to 10 years; and (3) the assumption that typical household size was no more than 4 to 5 people per family. Projections of prehistoric populations were closely tied to the presumed lack of durability of Mississpian-age, thatch-covered homes. Since they were constructed with walls of sapling poles set into foundation trenches, these rectangular structures would be prone to rot at their base within the frequently wet soils of the Mississippi River floodplain. The lack of structural integrity of these habitations therefore would have limited their effective use to as little as 5 years (Milner, 1998).

The urban core of Cahokia sustained between 3650 and 4000 individuals in a 10-km^2 area until approximately 700 BP, then underwent a period of demographic collapse as the American Bottom was abandoned over the next two centuries. The supporting satellite communities of the American Bottom were distributed across an area of 300 km^2 and supported 16 650 individuals by 900 BP, increasing to 25 000 people by 800 BP. This population declined to 5000 by 600 BP, after which people dispersed into the hinterlands of the uplands or other riparian settings. Abandonment of this portion of the vacant quarter was completed by 525 BP (Figures 7.6 and 7.7).

Archaeological evidence for resource utilization

The generalized subsistence base of Archaic hunter–foragers has been reconstructed based upon the kinds of faunal remains preserved in archaeological sites in the American Bottom (Kelly and Cross, 1984). For the past 4500 years, animal protein in the Indian diet was harvested from the rich array of habitats locally available. Faunal resources included a variety of mammals such as white-tailed deer, cottontail rabbits (*Sylvilagus*), and squirrels from terrestrial settings, in addition to 36 species of fish and four kinds of turtle caught in the active river channels, oxbow lakes and seasonally-flooded depressions, and 21 taxa of migrating waterfowl or resident birds. From Late Archaic through Mississippian times, the aquatic environment accounted for 80 percent of the individual prey animals identified from zoological remains, and the combination of semi-aquatic and terrestrial habitats comprised the remaining 20 percent. The reliance upon fish and seasonally migrating waterfowl along the Mississippi River flyway reflects both proximity and extent of wetlands. The area of open water in rivers and creeks, as well as in swamps and wet-meadow prairies, accounts for 35 percent of the natural features of the valley bottom (Milner, 1998).

The plant resources exploited by Native Americans were documented in conjunction with the interdisciplinary excavations of the Interstate 270 Project (Johannessen, 1984). Ethnobotanical remains were examined from 28 percent of all archeological features excavated, with plant debris recovered and identified from 23 500 liters of sediment, and tied by radiocarbon dates to 33 archaeological components of occupation (Johannessen, 1984; Rindos and Johannessen, 1991). Twenty-five summaries are available from floodplain sites for which plant materials were analyzed from features dating from 4400 to 500 BP (Figure 7.6). Eleven ethnobotanical summaries spanning the interval of 4750 to 850 BP include plant assemblages recovered from upland sites. The lack of upland sites post-dating 850 BP represents the absence of ethnobotanical information, despite the occurrence of some 100 upland archaeological sites dating from Mississippian times (Woods and Holley, 1991).

For the time span of the Late Archaic cultural period, 4750 to 3200 BP, and the Early Woodland period, 3200 to 2100 BP, wood charcoal recovered from the floodplain sites was nearly entirely of the kinds of tree species found growing in mid-to late-successional floodplain forests today (Zawacki and Hausfater, 1969). Of all the wood-charcoal fragments identified to taxon, more than 75 percent of the assemblage represented tree species forming closed forests on seasonally inundated floodplains. Johannessen (1984) defined the floodplain forest community as including species of elm, hackberry (*Celtis laevigata*), ash (*Fraxinus*), mulberry, and honey locust (*Gleditsia triacanthos*). The wood-charcoal assemblage dating 2350 BP from one site along the channel margin of the active Mississippi River was composed of 89 percent disturbance-favored trees of willow (*Salix*), cottonwood (*Populus*), river birch (*Betula nigra*), and sycamore (*Platanus occidentalis*). These species today form alluvial thickets on sandy point bars and stream edges. Upland archeological sites dating from 4500 to 2100 BP contained assemblages in which >80 percent of the wood charcoal was from upland trees such as oak and hickory. The charcoal material represents the combusted residue of fuelwood gathered by prehistoric people from the immediate vicinity of their residential areas (Johannessen, 1984).

From all sites, the high proportion of hickory nutshells indicates that prehistoric Native Americans preferred thick-shelled hickory as a dietary staple, prior to the cultivation of native herbs or the introduction of maize. Johannessen (1984) inferred a seasonal round of nut collection over a large territory, with favored nut resource patches dispersed from the floodplain with its trees of black walnut and thin-shelled pecan hickories (*Carya illinoiensis* and *C. cordiformis*), to upland stands for thick-shelled hickories (shagbark, pignut, and mockernut), acorns of both red and white oaks, and chestnut (*Castanea*), to upland prairie margins with ecotonal thickets of hazelnut (*Corylus americana*).

Edible starchy seeds with high carbohydrate content comprised 4 to 29 percent of all seeds recovered from floodplain samples dating to Early Woodland times (Figure 7.6). These seeds included goosefoot, knotweed, and maygrass, all domesticated members of the Eastern Agricultural Complex.

Archaeological evidence for progressive deforestation

During the Middle Woodland cultural period, from 2100 to 1600 BP, at archaeological sites within the alluvial valley, the wood-charcoal assemblages evidence a shift from use of floodplain species to those characteristic of upland environments (Figure 7.6). This change in selection of fuelwood has at least three plausible explanations: (1) increasing population sedentism and depletion of fuelwood from floodplain forests, requiring gathering of fuel from uplands; (2) cultural preference for the hotter and longer-burning hardwoods of oak and hickory for hearth use; or (3) clearance of garden plots or cultivated fields in the uplands and subsequent use of the cut-down trees for fuelwood (Johannessen, 1984).

A fourth explanation for the switch from bottomland to upland wood sources is that of the cumulative effects of burning for swidden agriculture. Clearance of bottomland forests would have opened light gaps through the canopy cover and would have offered successional opportunities for disturbance-favored weeds. The interpretation of local and progressive forest fragmentation is consistent with the increase in seeds in pit-storage caches and the gradual addition to the diet of herbs with edible, starchy seeds, including goosefoot, erect knotweed, and maygrass, supplemented by oily-seeded marsh elder and wild bean (*Strophostyles helvola*). Thus, with the development of gardens on both floodplains and upland sites in and near the American Bottom, mutualistic interactions of humans and plants in the Woodland cultural period led to domestication of native plants characteristic of the Eastern Agricultural Complex (Smith, 1992).

During the Late Woodland period, from 1600 to 1100 BP, concentrations of wood charcoal reached their maximum in archaeological sites from both the floodplain and the uplands. In all sites, wood charcoal of upland oak and hickory constitute >80 percent of all charcoal fragments. In the uplands, people might have pruned and culled less-productive trees in order for people to better curate hickory–nut orchards. High representation of upland oak–hickory charcoal in the bottomlands can be interpreted as the result of people bringing in hot-burning fuelwood. The relatively minor representation of floodplain trees in the Late Woodland reflects the elimination of local sources for fuelwood. People may have scavenged driftwood and regenerating woody brush in a lowland landscape that had been largely cleared for cultivation.

In the Late Woodland cultural period, expansion of riparian gardens is indicated by both substantially higher concentrations of small seeds, up to 90 percent of which were from plants of the Eastern Agricultural Complex. A decline in nutshell concentrations recovered from archaeological contexts may reflect a shift in the dietary role of hickory, a change in processing technique for more efficient extraction of nutmeats (Johannessen, 1984), or the displacement of less-productive forest stands by more intensive cultivation of starchy- or oily-seeded native herbs.

During the cultural periods of the Emergent Mississippian, 1100 to 900 BP, and Mississippian, 900 to 500 BP, maize supplemented the harvest of native starchy- and oily-seeded herbs (Figure 7.6). Analyses of stable carbon isotopes on human skeletons confirm that maize became a prominent component of the prehistoric human diet within the Central Mississippi Valley around 1000 BP, and remained so until the time of contact with European Americans (Lynott *et al.*, 1986; Greenlee, 1998). From 1100 to 850 BP, the proportional increase in less-palatable acorns to preferred hickory nutmeats may indicate the heightened pressure of exponentially increasing human populations upon diminishing foraging yields.

Wood-charcoal assemblages indicate a more diverse search for firewood in the Mississippian cultural period, with only 10 to 20 percent from floodplain forest species. In the interval from 950 to 600 BP, willow and cottonwood became important in the charcoal record. Both of these trees grow in clonal streamside thickets and are favored by episodes of flooding that scour the stream bed and then dump coarse sand in overbank deposits.

Such extreme flooding events would have differentially impacted harvests of cultivated herbaceous plants. Maygrass ripens in late May to early June; extreme floods occurring in summer would disrupt the crop of late-summer ripening goosefoot and knotweed, and fall-ripening maize (Johannessen, 1984). Increased severity and frequency of major floods would accentuate the risk of crop failure for these agrarian-based subsistence cultures. Milner (1998) documented that during the Late Mississippian cultural period, permanent Mississippian residences were displaced toward the highest ground on levee crests. Higher flood zones would intensify the competition for use of highest, driest sites that would have provided both a safe place to live and yet satisfied the need to ensure reliable harvests of high-yielding maize and seed crops.

Within the American Bottom, individual needs for selecting wood for hearths and home construction appear to have been over-ridden by broader governance decisions for defense of the communities. Lopinot and Woods (1993) presented the case for wood over-exploitation as a hypothesis to explain the collapse of Cahokia. Based upon the archaeological evidence for postmolds indicating the positions of wall posts, they projected that 800 000 timber posts were used for residences to house the prehistoric population. This represents a conservative underestimate, as it does not include the additional poles required for roof frames of the homes, public structures, and palisades, or the daily cooking and heating needs for fuelwood.

Starting in about 850 BP, the urban core of Cahokia was surrounded with a 3-km long stockade that served as a defensive barrier seven meters tall. This palisade wall apparently was built rapidly, cutting off outlying residential areas. Bastions placed at regular distances along the wall provided platforms from which archers could shoot arrows. Built and rebuilt four times between 850 and 650 BP, each construction event required a minimum of 15 000 logs of oak and hickory, with each pole 25 cm in diameter and about 8 meters long (Pfeiffer, 1974).

With witness-tree data from early-historic surveys, Lopinot and Woods (1993) projected that at most 628 000 trees could have grown in woodlands and savannas within a 10-km radius of the Cahokia center. Lopinot and Woods (1993, p. 214) speculated that because of earlier clearance of floodplain forests for cultivation, the landscape at 1000 BP "before the rise of Cahokia consisted of a mosaic of open woodlands, abandoned fields with even-aged stands of trees, and cultivated fields."

Based upon Mississippian wood-charcoal assemblages, 53 to 60 percent of the fuelwood probably came from nonlocal sources, and people needed to import wood from the uplands (Lopinot and Woods, 1993). With the population peaking by 800 BP, wood harvesting for palisades, homes, and hearths resulted in deforestation of the landscape within 10–15 km of Cahokia. Erosion of loess-mantled uplands would have meant increased transport of silt by upland streams. Resulting deposition of sediment in the American Bottom may have adversely impacted aquatic environments, decreasing the fish harvest. Denuded upland landscapes would have increased the flashiness and extent of overland flow of rainwater, compounding the risk of severe flooding of lowlands (Lopinot and Woods, 1993).

Geomorphic evidence for late-Holocene flooding

Holocene stratigraphic records from southwestern Wisconsin and northwestern Illinois, along with alluvial chronologies from the upper watersheds of the Mississippi and Missouri rivers, provide evidence for paleofloods as well as for the timing of major fluvial changes in flood frequency and magnitude over the past 11 500 calendar years (Knox, 1985, 1996). This region is hydrologically sensitive to minor climate change. It is situated across a major prairie/forest ecotone that is defined by steep climatic boundaries associated with seasonally shifting frontal zones of storm tracks. The typical annual flood regime determines the overall size of bankfull river channels and the elevation of aggrading floodplains. The flood regime is driven by late-winter snow cover, spring melting conditions, and the pulse of runoff from snowmelt. The Mississippi River generally aggraded through the late Holocene, between 5200 and 950 BP, with a stable hydrologic regime in which floodplain elevation rose gradually in response to regular flooding events. During certain time intervals, however, bankfull-stage flooding varied considerably from average modern conditions. Bankfull-stage floods were 10 to 25 percent smaller than today before 6800 BP, from 5200 to 3200 BP, and from 1600 to 1350 BP (Knox, 1996). In contrast, bankfull-stage floods exceeded modern-day floods by 10 to 25 percent during three relatively short episodes, from 6600 to 6000 BP, 1250 to 950 BP, and 50 to 30 BP. Historic forest clearance and cultivation, which increased sheetwash erosion and overland runoff, was responsible for the most recent interval of high floods (Knox, 1996).

In contrast to gradual trends in regular, long-term averages in stream behavior observed under stable hydrologic regimes, extreme flood events – including infrequent overbank floods of large magnitude – may destabilize the bottomland landscape. In the upper Mississippi River Valley, extreme floods are most commonly produced by intense, heavy summer rainfalls associated with stalled-out storm frontal systems. From 5200 to 3200 BP, the largest floods probably did not exceed the magnitude of a modern 50-year flood. However, the height of overbank floods, as well as the variability and magnitude of flood events, rose markedly at about 3200 BP. Further, a pronounced trend to more frequent occurrence of 500-year floods occurred from 650 to 400 BP, and these intense floods were stacked upon the peak elevations of typical floods that after 950 BP were some 25 percent higher than today (Knox, 1996). These flood events coincided with a pulse of increased frequency and severity of El Niño/La Niña precipitation events, documented by isotopic evidence to have occurred from 650 to 590 BP (Moy *et al.*, 2002).

The 4250 years of late-Holocene regular flood events from 5200 to 950 BP corresponds with the occupation of the American Bottom by increasingly sedentary populations of late Archaic, Woodland, and Early Mississippian people (Figure 7.6). The interval during which floodplain forests were cleared for cultivation, from 2100 to 1600 BP, coincided with the time of dropping flood levels. With flood elevations reaching 10 to 15 percent lower than today's levels from 1600 to 1350 BP, the maximum area was available for cultivation in the bottomlands along the Mississippi River.

The annual cycle of regular springtime flooding and nutrient-replenished alluvial plains favored the long-term cultural experimentation with domestication and cultivation of native seed-bearing weeds. Human clearance of forests may have temporarily destabilized river regimes, triggering their increased downcutting from 2100 to 1600 BP and inadvertently expanding the extent of arable bottomland. The aggrading river responded with renewed silt deposition from 1600 until about 950 BP.

Mississippian populations in the American Bottom, however, were vulnerable to high typical water levels and unprecedented extreme flooding events across the valley. At first, the population response was to abandon the lower levee slopes and establish homes on the highest ridge crests or to abandon the bottomlands for less risky upland sites. During the Late Mississippian interval, the regional climate-driven patterns of the bankfull and overbank flood regimes became more massive and frequent. Deforestation of floodplain sites by 1600 BP and upland sites by 950 BP accelerated upland erosion of silt into the bottomlands and would have amplified the magnitude and flashiness of severe flood events.

The emergence of Cahokia as a political chiefdom, market and urban center generated a local shortage of wood and caused the deforestation of uplands as much as 15 km from the American Bottom. The denuded floodplain and upland hillslopes set the stage for devastation during the series of catastrophic 500-year floods that occurred from 650 to 400 BP. Thus climatic, geologic, ecological, and anthropogenic interactions were all factors that forced the abandonment of Cahokia.

The case study from the American Bottom exemplifies how cultural over-exploitation of natural resources, dependence upon highly productive agricultural ecosystems, and heightened natural environmental variability could have all combined in prehistoric times to exceed thresholds of habitat continuity and geomorphic stability on a landscape scale. Rindos and Johanessen (1991) suggested that the widespread adoption and dependence upon maize after 1200 BP provided a new storable food source that sparked a population explosion but also played a key factor in the ultimate decline of the Mississippian lifeway. People moved to floodplain environments suitable for extensive cultivation of high-yielding maize crops. As population densities increased, so did their need for ever-increasing farmland and food reserves. Increased food production culturally augmented the carrying capacity of the landscape for humans. Clearing of nut groves of hickory and oak for upland gardens, however, eliminated a formerly key alternative food resource. The interaction of exponentially growing populations, and their mutualistic dependence on one maize crop of high although variable yield (compared with the former complex of starchy and oily seeds) heightened their vulnerability to catastrophic crop failure because of climatic variability shortening the growing season, infestation of maize by pathogens such as the corn borer, and rising water tables and summertime flooding destroying harvests. Relatively minor environmental perturbations would have generated severe disruptions of the maize agricultural ecosystem and the overshooting human populations sustained by it (Cohen, 1977). Summarizing the demographic consequences of a maize diet by Mississippian-age Native Americans, Rindos and Johannessen (1991, p. 45) contended that "The interaction of denser populations with an unstable subsistence base brought about increasingly common episodes of local food shortage. Humans, like other animals, respond to these times of stress simply by emigrating. Cahokia didn't collapse, it evaporated."

PREHISTORIC POPULATION DENSITIES AND LANDSCAPE THRESHOLDS

The ecological potential for human-generated impacts is tied to the duration and continuity of aboriginal occupation, human population densities, and technological capabilities associated with subsistence-based lifestyles. During the last thousand years, many populations used maize agriculture as their primary subsistence base or to supplement other native crops. Part of agrarian communities, these people represent the third panarchical level of human ecosystems, that of chiefdom agriculturalists (Figure 3.5).

Within the Great Lakes and New England regions, native people lived at the northernmost agricultural limit, where killing frosts in late spring and early autumn frequently shortened the frost-free growing season to less than 120 days, the minimum growing season necessary for primitive races of maize (Cronon, 1983; McAndrews, 1988). In southern Ontario, Iroquois communities established small settlements, with at least eight phases of agricultural clearing documented within a 5-km radius of Crawford Lake between 650 and 250 BP. Archaeological excavations from 11 separate sites permit population estimates averaging 900 residents for each prehistoric community (Byrne and Finlayson, 1998) (Table 7.1). Annually laminated sediments from Crawford Lake record a fine-scale history of forest clearance, with evidence of human-set fire and cultivation of corn. Ephemeral occupation of villages is recorded in archaeological sites located within a 2-km radius of the paleoecological site. Even if only one village community averaging 900 members moved from one farming site to another "virgin site" during this 400-year interval, requiring a shift in location of settlements every 50 years or so, the archaeological record would consist of many past village sites. Within a 5-km radius extending from Crawford Lake, a prehistoric population of 900 residing in the overall 78.5 km^2 of catchment area translates to an overall density of 11.5 individuals per km^2 (Table 7.1). Within concentric rings of human impact concentrated within a more restricted 2-km radius of their village center, the long-term impact would have been a cumulative patchwork of fragmented forests, canopy light gaps cleared for crops, and successional regrowth of disturbance-favored forest. This total impact zone of about 138 km^2 is nearly 1.8 times the present catchment area, representing a disturbance rotation interval of roughly 230 years. The first two centuries of human presence would have been sufficient to convert the initial old-growth beech–maple forest into successional stands of white pine and northern red oak, documented in the paleoecological sequence and by the oldest remaining forest stands (Byrne and Finlayson, 1998). Thus, small groups living and farming in spatially separate locales for several centuries, then abandoning the area, left an ecological overprint of fundamentally altered forest communities.

Prehistoric populations living on the alluvial terraces of the Little Tennessee River in eastern Tennessee numbered between 2200 and 2400 in

Table 7.1 *Prehistoric human population densities at three locations in eastern North America during the Mississippian cultural period*

Location	Occupation interval (BP)	Archaeological estimate of size for prehistoric populations	Resettlement interval for shifting population residences	Area (km^2)	Population density (no./km^2)
Iroquois Townhouse villages near Crawford Lake, southern Ontario (Byrne and Finlayson, 1998)	650–250	900	50-yr rotation with eight phases of forest clearing	~78.5	~11.5
Little Tennessee River Valley, eastern Tennessee (Baden, 1987)	1000–250	2200–2400	125-yr rotation		
• Cultivated alluvial bottoms				~42.5	~52 to 56
• Whole landscape watershed				~140	~16 to 17
Mississippi River Valley, southern Illinois (Milner, 1998)	1000–525				
• Cahokia metropolis		4000	Fixed location	10	400
• American Bottom agricultural corridor		25 000	Fixed location	300	83

Late Mississippian and Early Historic times (Baden, 1987). Native Americans continuously occupied the rich alluvial bottoms for at least the past 4000 years. From 1000 to 250 BP, population density was about 52 people per km^2 (Table 7.1). The alluvial flats comprised some 30 percent of the broader watershed. Including the uplands paralleling 5 km either side of the Little Tennessee River, Native American use of the uplands for gathering fuelwood, cutting wood timbers, and collecting mast nuts would have dispersed their impact across a wider zone of habitats. Thus the effective population density over the whole watershed in preColumbian times was 2200 people per 140 km^2, or about 16 people per km^2 (Table 7.1).

In the American Bottom of southern Illinois, along the Mississippi River Valley, the chiefdom of Cahokia established from 1000 to 525 BP was supported by maize agriculture as well as by supplemental food reserves received as tribute to the reigning chiefdom rulers. With 4000 permanent residents concentrated within 10 km^2, Cahokia supported some 400 people per km^2 (Table 7.1). Milner's (1998) projections for the linear agricultural corridor indicate a resident base of 25 000 members, dwelling in the 300 km^2 of floodplain and upland borders. Thus the prehistoric population of the American Bottom as a whole was some 83 people per km^2 (Table 7.1).

CONCLUSIONS: PREHISTORIC SEDENTARY AGRICULTURE AND PANARCHICAL COLLAPSE

The ecological consequences of prehistoric sedentary agriculture in the Eastern Woodlands were manifold. Even at the northern limits of maize agriculture, and even with relatively low population densities, shifting agriculturalists who moved to new cultivated sites on a resettlement interval of as little as 50 years caused a landscape-level shift in forest dynamics. Within the time span of several centuries, natural ecosystems of late-successional, "virgin" forests were culturally transformed into an early-successional, anthropogenically managed mosaic of forests and old-fields.

When people became tethered to fixed sites for more than two centuries and when population densities exceeded 50 people per km^2 of cultivated land, cultural activities resulted in environmental degradation as native forest ecosystems became fragmented beyond the connectivity threshold. In the American Bottom of the Central Mississippi River Valley, as well as in the Little Tennessee River Valley, this critical ecological threshold was reached in Late Woodland and Emergent Mississippian times. When coupled with climatic and geomorphic changes that destabilized the rivers along which prehistoric people were concentrated, the landscape shifted from a stable mode to an unstable mode that resulted in unsustainable living environments. The resulting panarchical collapse (Figure 3.5) was part and parcel of the prehistoric cultural instability that led to evacuation of the Mississippian heartland a little more than 500 years ago (Figure 7.1).

8

Regional-level interactions

PALEOINDIANS, MEGAFAUNAL EXTINCTIONS, AND PANARCHY REORGANIZATION

During the late Pleistocene, humans interacted with their environment on a regional scale during a time of major climate change and ecosystem reorganization. Human predation, climate change, plant–animal interactions, and disease each may have played a role in the collapse and reorganization of regional ecosystems at the end of the Pleistocene (Figure 3.4). Regional-level influences of Paleoindians followed from success in immigration, establishment, and increase in human populations to a threshold at which they began to play a discernible role as agents of ecological change in the New World. Adaptations for hunting and foraging (Figure 3.5) would have influenced not only the regional dispersal pattern of human populations across the Americas at the end of the Pleistocene, but also would have determined the extent to which Paleoindians would have been active participants as keystone predators in the extinction of megafauna (adult mammals >44 kg).

DYNAMICS OF LATE-PLEISTOCENE ENVIRONMENTAL CHANGE

During the late Quaternary, mean global temperature warmed by 5 °C from peak glacial cold 18 000 BP to peak interglacial warmth 6000 BP Climate warming was accompanied by changes in hemispheric patterns of atmospheric circulation and shifts in regional storm tracks (Wright, 1989; Lowe *et al.*, 1994). Seasonal temperature contrast at middle and high latitudes heightened after 17 000 BP and peaked between 12 000 and 9000 BP (Figure 8.1) as systematic changes in incoming solar radiation forced the changeover from cool, equable ice-age climate to the highly seasonal late-glacial and early-interglacial pattern of very cold winters offset by very warm summers (Kutzbach and Guetter, 1986).

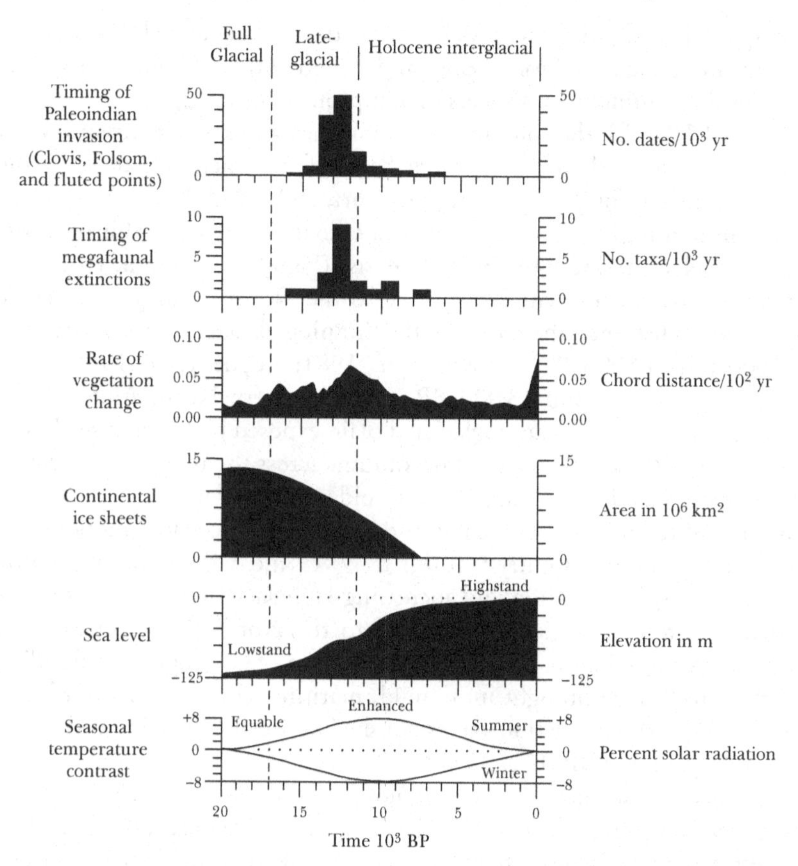

Figure 8.1 Late-Pleistocene and Holocene panarchical reorganization associated with environmental change, disassembly and reassembly of terrestrial plant and mammal communities, and invasion of North America by Paleoindians (modified from Bloom, 1971; Meltzer and Mead, 1983; Kutzbach and Guetter, 1986; Jacobson *et al.*, 1987; Bonnischen *et al.*, 1987; and Grayson, 1991).

The oxygen isotope record from cores of the Greenland Ice Sheet registers the glacial–interglacial transition as a sequence of stair-step shifts in temperature between 14 670 and 12 900 BP and after 11 650 BP. Interglacial temperatures were reached 10 000 BP (Stuiver *et al.*, 1995; Stuiver and Grootes, 2000). At the continental scale, these "flickering switches" were a series of partial changeovers between two fundamentally different atmospheric circulation patterns: (1) an ice-age mode with a jet stream split into two branches deflected around the northern and southern perimeters of the merged Cordilleran–Laurentide ice sheets; and (2) an ice-free, interglacial mode with one prevailing jet stream that steered mid-latitude storms along the Polar Frontal Zone and shifted northward during deglaciation of the continental ice sheets (Thompson *et al.*, 1993).

After 17 000 BP, the combination of longer, warmer summers and warm rainfall transported to the southern flank of the glacial ice mass by the southern jet stream (Delcourt and Delcourt, 1984, 1993) accelerated summertime melting of ice, returned meltwater to the oceans, destabilized floating ice shelves, and resulted in sea-level rise (Figure 8.1; Bloom, 1971). Ice streams carried continental glacial ice to the North Atlantic and Pacific Oceans and hastened the retreat of the thinning glacial ice masses between 17 000 and 11 500 BP (Mayewski *et al.*, 1981; Hughes *et al.*, 1985).

Between 40 000 and 13 000 BP in northwestern North America, the Bering land bridge was an unglaciated plain exposed by lowered sea levels. During the Wisconsinan glacial maximum, across the interior of Beringia the climate was drier and appreciably colder than today. After 14 700 BP, the annual temperature range increased regionally, exceeding modern values for maximum summer warmth by 7 °C and winter minimums by 5 °C. High solar radiation and increasing late-glacial temperatures accelerated the return of glacial meltwater to the North Pacific Ocean and inundated coastal areas of the Bering platform. The progressive flooding of the land bridge brought more mild, maritime climate northward into Arctic Beringia. Circulation was established between the Pacific and Arctic Oceans by 13 000 BP (Elias, 2000).

Perched at the shelf edge of the northwest Pacific Coast, present-day island chains such as the Alaskan Alexander Archipelago and the Canadian Queen Charlotte Islands offered ice-age refugial areas largely free of glacial ice (Warner *et al.*, 1982; Barnosky *et al.*, 1987; Heusser, 1989). At its maximum extent between about 24 000 and 18 600 BP, the Cordilleran Ice Sheet fed outlet glaciers flowing west to the Pacific Ocean. After 17 000 BP, however, calving of icebergs caused the glacial margins to retreat eastward, as open-water fjords inundated the glacially scoured troughs. Accompanying deglaciation between 17 000 and 14 600 BP, rapid isostatic rebound of the land lowered the relative position of sea level by as much as 150 m. Exposed by seaward retreat of the uplifted terrain, the exposed coastal plain was broken by fjords up to 20 km across. Rising sea level once again overtopped uplifting landscapes after 11 500 BP, as the early-Holocene transgression expanded open-water distances among stranded island refugia (Barrie and Conway, 1999; Mandryk *et al.*, 2001). Thus, the Pacific Coast was an open-water corridor available for human colonization both before 24 000 BP and after 17 000 BP.

An Ice-free Corridor between the Cordilleran and Laurentide ice sheets extended through west-central Canada during the late Pleistocene. Geological investigations indicate that a north–south gateway was open between the Cordilleran and Laurentide ice sheets from 55 000 to 26 000 BP, after which this gateway was closed off by merging glacial-ice flow (Dyke and Prest, 1987; Wright, 1991). The Ice-free Corridor reopened at approximately 15 600 BP and was at that time as narrow as 25 km wide in places, strewn with unstable remnant ice blocks and interrupted by torrents of meltwater that fed braided streams and glacial-margin lakes (Haynes, 2000). Widening of the corridor may have been delayed until as late as 13 000 BP (Wilson and Burns, 1999; Mandryk *et al.*, 2001).

From 24 000 to 15 600 BP, ice domes on land masses bordering the North Atlantic Ocean extended seaward as a persistent and laterally continuous ice shelf, providing over-ice access bridging from northwestern Europe, to Iceland, Greenland, and northeastern North America (Ruddiman and McIntyre, 1981; Imbrie *et al.*, 1983; Delcourt and Delcourt, 1984; Ruddiman, 1987). North of the full-glacial position of the maritime polar front (Figure 8.2), winter pack ice formed south to the Iberian Peninsula, spanned the North Atlantic Crescent at 48° N, and joined the American Delmarva Peninsula at about 40° N. With summertime warming, ice floes formed as the pack-ice margin retreated back to the permanent ice shelf as far as 60° N near Greenland. Between 15 600 and 13 000 BP, the climatic boundary of the Polar Front Jet Stream deflected northward to as much as 60° N (Iceland), and warm, saline surface waters flowed as the North Atlantic Drift, substantially reducing seasonal formation of sea-ice offshore of northwestern Europe (Figure 8.2; Ruddiman and McIntyre, 1981; Ruddiman, 1987).

LATE-GLACIAL RESTRUCTURING OF ECOSYSTEMS

Plant communities

During peak glacial time, across the middle latitudes of the unglaciated United States, the southern jet stream anchored the ecotone between boreal-like coniferous forest and temperate mixed deciduous–conifer forest at about 34° N (Delcourt and Delcourt, 1984, 1993; Figure 8.3). In the southeastern United States, the greatest rates of change in plant community composition in the late-glacial interval occurred with initial climate warming between 17 000 and 14 700 BP (Delcourt and Delcourt, 1984, 1987). In the deglaciated Midwest and New England, fastest vegetational

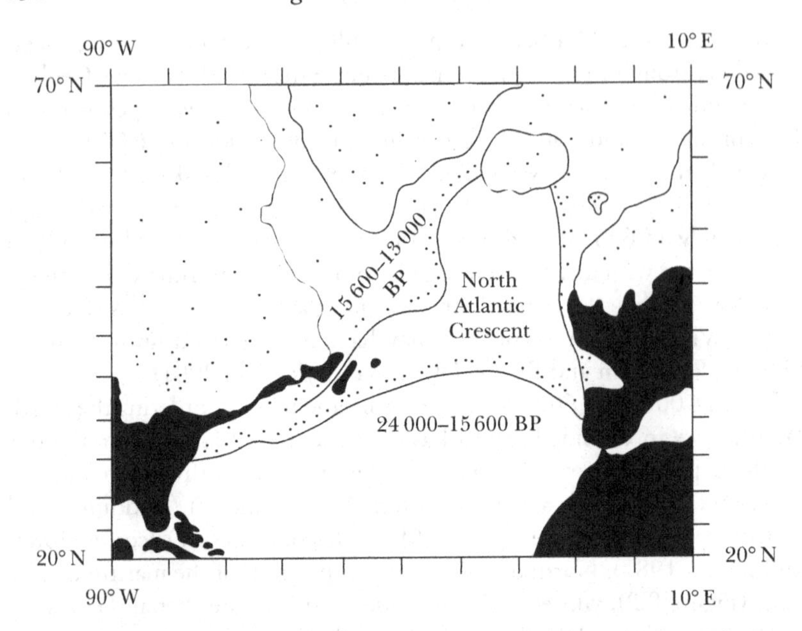

Figure 8.2 The North Atlantic Crescent, stepwise shifts in the southern edge of the ice shelf and marine pack-ice in the North Atlantic Ocean between its full-glacial and early late-glacial position (24 000 to 15 600 BP) and a more northern late-glacial position (15 600 to 13 000 BP). The full-glacial sheet of winter pack-ice extended south to 45° N, the maritime position of the Polar Jet Stream and the warm Gulf Stream current (adapted from Ruddiman and McIntyre, 1981; and Ruddiman, 1987).

changes occurred at the Pleistocene/Holocene transition from 13 500 to 10 200 BP (Overpeck *et al.*, 1985; Jacobson *et al.*, 1987) (Figure 8.3). Shifts in major vegetational boundaries were time-transgressive, moving from south to north from 17 000 to 11 500 BP, and from low to high elevations in the Appalachian Mountains between 15 600 and 11 500 BP (Delcourt and Delcourt, 1998b). In eastern North America, peak migration rates for cool–temperate tree species were from 17 000 to 14 000 BP. Boreal trees previously blocked by barriers such as slowly retreating glacial ice spread northward at their fastest rates between 14 000 and 9000 BP (Delcourt and Delcourt, 1987).

In response to Milankovitch-cycle inputs of solar radiation that magnified winter–summer temperature differences starting 17 000 BP, plant communities disassembled and reassembled between 14 000 and 11 500 BP at middle latitudes, and from 13 000 to 7800 BP across higher latitudes of deglaciated North America (Delcourt and Delcourt, 1987, 1994). Because of their individualistic and opportunistic responses to variable and dynamically changing weather and disturbance conditions, late-glacial and

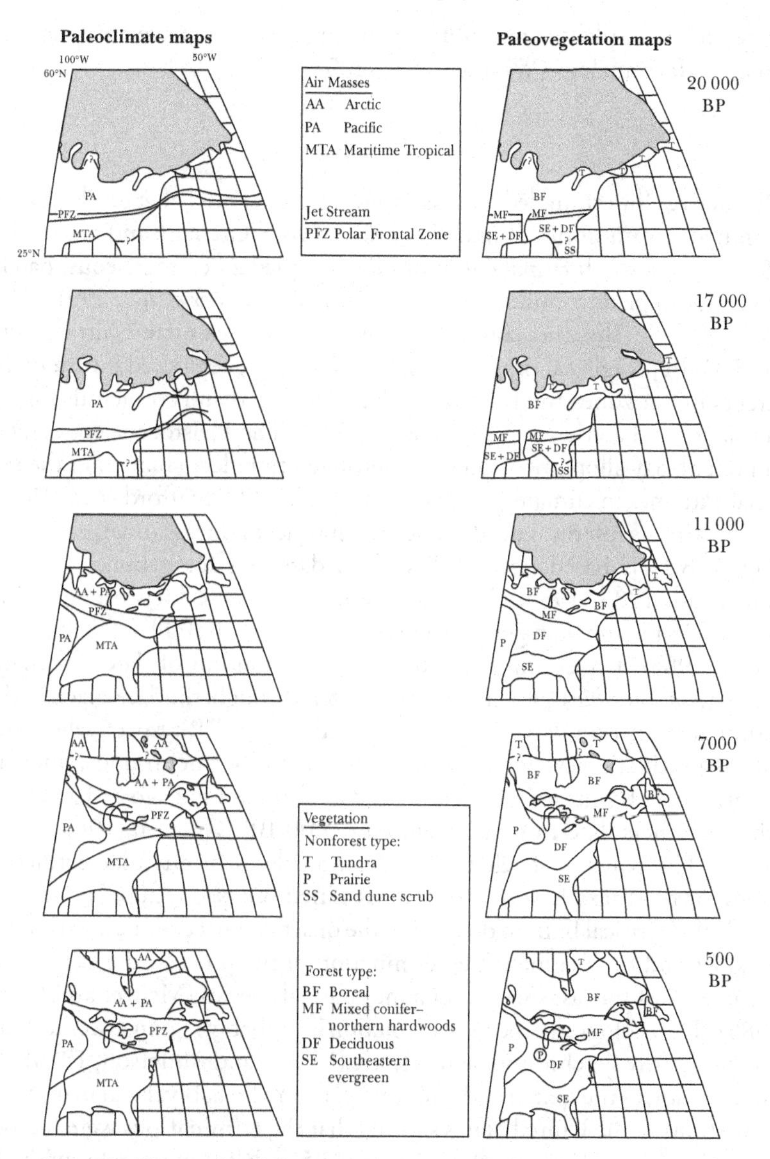

Figure 8.3 Glacial–interglacial changes in past climate, areal extent of glacial ice (shaded areas), and vegetation mapped across eastern North America for the last 20 000 calendar years. The maps reflect full-glacial times (20 000 BP), the late-glacial interval (17 000 BP), and the Holocene interglacial (11 500, 7000, and 500 BP). Paleoclimate maps identify climate regions dominated by Arctic, Pacific, and Maritime Tropical air masses. The typical path of the Polar Jet Stream corresponds with the Polar Frontal Zone (modified from Delcourt and Delcourt, 1987).

early-Holocene plant communities were ephemeral and have only relatively poor modern analogs (Williams *et al.*, 2001).

Mammal communities

Distinctive "intermingled faunas" were characteristic of late-Pleistocene mammal communities (Lundelius *et al.*, 1983; Graham and Lundelius, 1984). Species-rich faunal communities occupied a heterogeneous, patchy landscape of plant communities (Guthrie, 1984; Turner *et al.*, 1999) within cool, equable climatic conditions with modest seasonal contrast (Figure 8.1). In postglacial times, terrestrial faunas have separated into very different climate zones and habitats, such as deciduous and coniferous forests, grassland, and arctic tundra (Graham, 1986). The Pleistocene coexistence of present-day allopatric species is interpreted to reflect changes in the seasonal extremes in climate (Graham and Mead, 1987; Stafford *et al.*, 1999).

The timing of the late-Pleistocene episode of megafaunal extinction (Figure 8.1) is based upon radiocarbon dates on fossil bones (Meltzer and Mead, 1983, 1985; Grayson, 1989, 1991). Late-Pleistocene extinctions of 25 mammalian genera were completed by 11 500 BP (Meltzer and Mead, 1985). For example, the temporal distribution of dates for mammoth (*Mammuthus*) and mastodon cluster through the late-glacial with terminal extinction dates of 12 750 BP and 12 200 BP, respectively. However, the extinction event may not have been broadly synchronous, since last occurrences of some extinct mammals date from as long ago as 15 600 BP, whereas others date from as recently as 7000 BP (Grayson, 1989, 1991, 2003). The paucity of reliable radiocarbon dates means that extinction times for most taxa have yet to be determined (Grayson, 2003).

It is thus critical both to determine the onset and pace of these extinction episodes and to document the termination of the process, the last known occurrence of the last survivor of a megafaunal species (Meltzer and Mead, 1985). The ecological process of extinction is probably region-specific and is not necessarily synchronous across space (Kurtén and Anderson, 1980). An examination of the spatial and temporal pattern of reliably dated megafauna extinctions in the United States showed that the terminations were focused in the brief time window of 14 000 to 11 500 BP (Figure 8.1), with the geographic pattern of megafaunal extinction moving from the southeast to the northwest across the United States (Beck, 1996).

Climate change may have been the ultimate factor accounting for megafaunal extinction (Grayson, 1991), with additional factors also contributing. The southeast-to-northwest pattern of proboscidean extinction is spatially consistent with an interpretation of increasing "environmental insularity" (King and Saunders, 1984). The spatial continuity of habitats preferred by American mastodon (*Mammut americanum*) populations was disrupted by fragmentation of late-Pleistocene spruce and jack pine forest between 33° and 40° N latitude (Figure 8.3). Proboscideans became concentrated in these geographically restricted and environmentally insular patches of conifer forests, where they were progressively isolated by an expanding matrix of temperate deciduous forest. Deterioration and increased patchiness of island-like conifer stands would have reduced the area of habitat necessary for continued existence of mastodons. A decline in reproduction, combined with stochastic events, could have led to their extinction (King and Saunders, 1984).

During the late Pleistocene, herds of Columbian mammoth (*Mammuthus columbi*) grazed in open woodlands and grasslands (Graham and Lundelius, 1994). In late-glacial times, their populations may have increased markedly in size despite predation pressure, as they migrated northward into Canada to occupy an expanded range in highly productive pioneer "pulse" ecosystems that were assembling on nutrient-rich deglaciated landscapes (Geist, 1999).

Ecological bottleneck for mammal communities

The late-glacial disassembly of intermingled mammalian faunas may have been primarily the product of the ecological bottleneck imposed by the glacial–interglacial maximum in seasonal contrast and shift to interglacial warmth between 17 000 and 11 500 BP. These profound climate changes altered the physical character and spatial extent of terrestrial environments available for colonization, and "caused individual species distributions to change along environmental gradients in different directions, at different rates, and during different times" (Graham, 1990, p. 58).

At the Pleistocene/Holocene transition, the spectrum of plant–herbivore interactions was altered as communities of megaherbivores changed in composition and geographic extent, and as species became extinct. After this panarchical collapse, new ecosystems emerged within the early Holocene. Guthrie (1984) used an analogy of plaids and stripes to describe the transformation of a relatively fine-grained mosaic of glacial-age vegetation (plaid) to a relatively coarse-grained mosaic of interglacial vegetation (striped) across much of boreal and arctic North America. The loss of vegetation heterogeneity severely impacted diversity of habitat and of potential plant foods available for small mammals with limited ranges, forcing local faunal extinctions with consolidation of more contiguous, core distributions. Stress-tolerant, evergreen spruce trees and nitrogen-fixing alder shrubs, which produce biochemical defenses against herbivory, dominated many coarse-grained patches of boreal vegetation, reducing the nutritional status of plant food and effectively lowering the carrying capacity for megaherbivore populations (Guthrie, 1984; Graham, 1990).

The late-Pleistocene conjunction of a diverse guild of megaherbivores, massive environmental change, and deteriorating food base may have significantly magnified competition among herbivore populations, enhancing their vulnerability to extinction (Lambert and Holling, 1998). Mammals with adults of large body size, fragmented distribution range, ecological specialization, and with environmental and reproductive isolation may be considered as "insular endemics" with high probabilities of extinction (Brown, 1995).

Late-glacial conditions of extreme winter cold and deep snow-cover would further have stressed populations of large herbivores by limiting winter forage (Guthrie, 1984; Graham, 1990). Extended freeze-over of water sources would complicate this seasonal crisis of severe water shortage for mammoths, as effectively as summertime droughts constrict elephant movements today to remaining water refuges (G. Haynes, 1991; Holliday, 2000). Environmental stress, resulting from warming, drying, and harsh seasonal extremes, may have led to density-dependent responses with periodic crowding of megaherbivores around remaining waterholes, overgrazing and depletion of nearby plant-food resources, increasing combative competition for finite water supplies, differentially culling juveniles and self-regulating new births, and predictably impacting herd longevity and vulnerability because of individuals' injury, disease, or opportunistic exploitation by keystone predators such as Clovis hunters (C. Haynes, 1991, 1995).

ALTERNATIVE ECOLOGICAL MODELS FOR RAPID COLONIZATION OF NORTH AMERICA

Archaeological evidence for oldest human presence predates widespread colonization of the Americas by Clovis people around 13 500 BP. PreClovis artifacts, for example those preserved at Monte Verde, Chile, in sediments dated securely to 14 700 and tentatively to more than 35 000 BP (Dillehay, 1997, 2000), were left behind by unsuccessful colonists. Earliest human visitors to the North American continent made landfall in geographically separate areas, and their arrival was in most cases probably followed by calamity. Populations failed through excessive mortality, genetic bottlenecks in founder groups, or insufficient reproductive success, leading to extinction. These failed efforts in colonization were the result of populations becoming reduced below sustainable thresholds. A significant and persistent human presence required sustained colonization of immigrants in one or more pulses or a continuous "dribble" (Meltzer, 1993), followed by a sustained population increase. In contrast, hunter–foragers of the terminal Pleistocene were highly successful. With generalized tool kits, lack of competition, and mobile foraging practices, their colonizing strategies were unlikely to have been random.

Two plausible cultural strategies for successful colonization of the New World are those of transient explorers and estate settlers (Beaton, 1991). These strategies differ in style and potential magnitude of ecological impact, employing separate methodological logic, foraging practices, and demographic patterns for growing populations and splintering off new tribal units. Transient explorers would have budded off in dispersal units as small as bands consisting of minimal family units of a man and a woman. They would have traveled long distances to new locations and habitats. Populations of transient explorers would have remained small and isolated from one another, leading to low fecundity and high inbreeding. Stochastic events would have resulted in high local extinction rates. Food resources would have been selected as a compromise between search time (resource distribution) and pursuit time (successful kill or harvest). These highly mobile foragers and pursuers of game may have selected for favored habitat types, gleaning readily obtainable food, and then moving on. Using a generalized tool kit, they might have relocated in new ecological settings, dispersing rapidly as small bands across large areas (Beaton, 1991).

The estate settlers' strategy would have involved a more conservative and sedentary approach to the peopling of the New World. In the estate settlers' strategy, immigrants would have established a home territory, or estate, within a familiar ecological context of habitats and food options. Population growth, infrequent fissioning off of new bands consisting of many individuals, and maximum food extraction from a limited collecting area would have ensured the colonists' relatively high fecundity, large number of breeding partners, and lower extinction rates than experienced by transient explorers. Estate settlers would have utilized a more varied diet, combining hunting and foraging in a variety of habitats within their home territory. Successful colonization of the North American continent by estate settlers would have been tied to their selection of "mega-patches" or ecoregions in which humans had already attained demographic success – first founding new estate colonies along marine coastlines, then extending inland along river floodplains (Beaton, 1991).

Five spatially explicit models of Paleoindian colonization have been proposed to reconcile the archaeological evidence with plausible scenarios driven by terrain or by biome-level opportunities afforded the earliest nomads. Four of these models are compatible with transient explorers and rapid continental spread: (1) the wave-front diffusion model of Overkill (Martin, 1967, 1984; Mosimann and Martin, 1975); (2) the least-cost pathways model (Anderson and Gillam, 2000); (3) the off-the-shelf hypothesis (Anderson and Faught, 1998), and (4) the Solutrean solution (Stanford and Bradley, 2000). The fifth hypothesis, the megapatch model (Dixon, 1999) exemplifies the regional expression of estate settlers. These scenarios incorporate presumptions for the geographic entry and timing for arriving founder populations.

Wave-front diffusion model of Overkill

This scenario is predicated on the passage of Clovis hunters from Beringia through the interior route of the Ice-free Corridor, arriving near Edmonton, Alberta in 13 500 BP (Martin, 1958, 1967). The model proposes that once beyond the glacial-ice barrier, populations of these nomadic hunters expanded along a bow-shaped arc, as a population wave front diffusing across the Americas independent of terrain constraints. Impervious to topographic barriers, these nomadic tribes would have perceived their New World as isotropic, with southerly radiations along a broad spectrum of equally attractive paths.

In the late Pleistocene, as proposed by Mosimann and Martin (1975, p. 304):

> Paleolithic invaders tracked herds of prey eastward through unglaciated parts of Alaska to an ice-free corridor that led east of the Canadian Rockies and south to the heart of the continent. These people came from a long lineage of skillful hunters with hundreds of thousands of years of Paleolithic experience behind them. They were expert at tracking, killing, butchering, and preserving meat from large mammals. It appears that they ate little else.

The leading edge of migration was thus envisioned as a front consisting of a high-density population that was responsible for intensive hunting of large mammals and rapid advance of the wave of human immigrants. As the human population grew at a geometric rate, with no limitation on available resources to restrict its intrinsic rate of increase, expansion into previously uninhabited territory would have taken place at an ever-increasing rate. Only when prey was depleted in an area would humans move long distances, causing the wave front to "jump" into new territories (Whittington and Dyke, 1984). With demographic simulations over a 1000-year span, Mosimann and Martin (1975) projected the wholesale elimination of late-Pleistocene megafauna, as nomadic hunters traveled from the Alberta Ice-free Corridor to the southernmost tip of South America.

The Overkill hypothesis linked Early Paleoindian arrival in the New World and late-Pleistocene extinction of large mammal species (Figure 8.1). The Overkill model proposed a causal relationship between Paleoindian arrival and Pleistocene extinctions, based on several key observations and assumptions: (1) the extinction event was dated within a narrow chronological range *c.* 13 000 BP; (2) loss of species was concentrated within large-bodied mammals; (3) species were eliminated without being replaced by ecological counterparts or evolutionary descendents (G. Haynes, 1991); and (4) evidence for major climatic change purportedly did not coincide with the time of the extinction event (Meltzer and Mead, 1983, 1985).

Mosimann and Martin (1975) proposed that Paleoindian hunters harvested populations of large mammalian herbivores to extinction. As an ecological consequence, those creatures dependent upon herbivores, their carnivore predators, scavengers, and commensals, experienced a cascading suite of extinctions following the collapse of their herbivore food base (Grayson, 1989). According to Martin (1963, p. 70): "large mammals disappeared not because they lost their food supply, but because they became one."

Butchered remains of only two megafauna genera, mammoth and mastodon, have been commonly recovered from documented archaeological "kill sites" (Grayson, 1991), while other potentially suitable prey, including camel (*Camelops*), tapir (*Tapirus*), peccary (*Platygonus*), and horse (*Equus*), were equally or more abundantly represented from late-Pleistocene vertebrate sites (Lundelius *et al.*, 1983; Grayson, 2002). Based on this observation, Mosimann and Martin (1975) simulated predator–prey interactions presuming a dietary preference for ancient proboscideans. Hunting by Paleoindians, however, may have included a broader spectrum of game animals as opportunity availed. For example, in the eastern Beringian sites of Alaska and the Yukon, dried blood residues have been analyzed from 16 fluted projectile points, recording the Paleoindian hunting and butchering of a now-extinct species of mammoth (*Mammuthus primigenius*). This analysis confirmed that mammoth kills were contemporaneous with hunting of species that are still extant, including caribou, grizzly bear (*Ursus arctos*), bison, musk ox (*Ovibos moshatus*), and mountain sheep (*Ovis dalli*) (Loy and Dixon, 1998).

Least-cost pathways model

In contrast with the Overkill model for terrain-independent saturation of human colonists, other models have emphasized the anisotropic character of terrain in potentially funneling nomadic invaders along preferential corridors of relatively easy access. Least-cost models for migration were developed by Anderson and Gillam (2000, 2001), predicated on two observations: (1) topography, habitats, and plant and animal resources in the late Pleistocene were not uniformly distributed throughout the North American continent; and (2) many of the largest concentrations of Clovis points are found in the southeastern United States.

Scenarios were developed for two entry routes: (1) the interior Ice-free Corridor; and (2) the Pacific Coastal Corridor (Fladmark, 1979, 1990). The latter was made available as late-glacial retreat of the Cordilleran Ice Sheet and seaward retreat of isostatically uplifted terrain exposed a potential immigration corridor between 17 250 and 11 500 BP. This coastal pathway closed when sea level rose, inundating the exposed coastal strip and expanding open-water distances among stranded island refugia (Barrie and Conway, 1999; Mandryk *et al.*, 2001).

Topography may have guided the coastal and interior routes selected by Paleoindians (Anderson and Gillam, 2000, 2001). The rapid increase in archaeological sites post-dating 14 000 BP may reflect the dramatic expansion of Paleoindian populations shortly after their initial entry (Figure 8.1). Physical barriers, such as ice sheets, great freshwater lakes, and mountain ranges may have redirected movement of exploring bands, funneling nomads toward easier pathways of lowland plains, coastal shorelines, glacial-lake margins, and river floodplains. By comparing fine-scale topography with the locations of 44 key early-human archaeological sites, Anderson and Gillam (2000) projected several "least-cost pathways" for migration. As determined by terrain analysis, all migration pathways reflected the steering role of local topography, which funneled the land-based travelers along coastal and major river systems. Simulations included estimates of population size of founding groups, population growth rate, group size, and optimal foraging area for nomadic bands. Late-Pleistocene colonizers were projected to have dispersed throughout the Americas within one thousand to several thousand years of their entry.

Portrayed as settlements with circular group "territories" (400-km diameter), Anderson and Gillam (2000) speculatively presented the succession of territories invaded then abandoned by Paleoindians as vital resources were exhausted. With their entry from the north via Beringia or via the coastline of the Pacific Northwest, plausible least-cost routes for Amerindian colonization can reflect the successive occupation of new territories (shown as "pearls"). These plausible scenarios reflect iterative steps in southward migration to new staging areas, differing primarily in choice of either short dispersal jumps ("string-of-pearls" model; Figure 8.4) or long-distance movements to new resource-rich locales ("leapfrog" model) (Anderson and Gillam, 2000).

The string-of-pearls model is consistent with spatial diffusion models. These models show that for invading species of plants and animals, the more foci for initial colonization, the faster the rate of range expansion (Mack, 1985). Each additional focal point for colonization accelerates the process of species invasion, even if only the number of foci, and not the total area initially occupied, is increased. Spread of an invading species population is also enhanced by foci that are far apart, because more area can be occupied before population ranges coalesce. Repeated invasions may have similar results to multiple simultaneously established foci in enhancing the spread of a population, provided that the introduction of the population is not made repeatedly at the same point of initial entry (Mack, 1985).

The possibility of "leap-frog" or jump dispersal as a mechanism for rapid advance of immigrating human populations is supported by data from dispersal curves of other species of invading organisms (Kot *et al.*, 1996). Data on dispersal of a wide variety of organisms, from insects and birds to seeds and pollen of vascular plants, show that the spatial distribution of propagules about a source is typically leptokurtic, with more offspring or propagules dispersed near the center and in the tails of the distribution than in a normal distribution of comparable mean and variance. Mathematical models incorporating leptokurtic distributions show that the speed of invasion of a spreading population may be accelerated over that expected in a traveling wave of normal distribution because of success of a relatively small number of widely dispersed individuals (Kot *et al.*, 1996).

Off-the-shelf hypothesis

The "off-the-shelf" hypothesis of Anderson and Faught (1998) proposes spatial gradients in human impact, driven by their initial exploration as anchored to the coastal zone of maximum ecological diversity for subsistence,

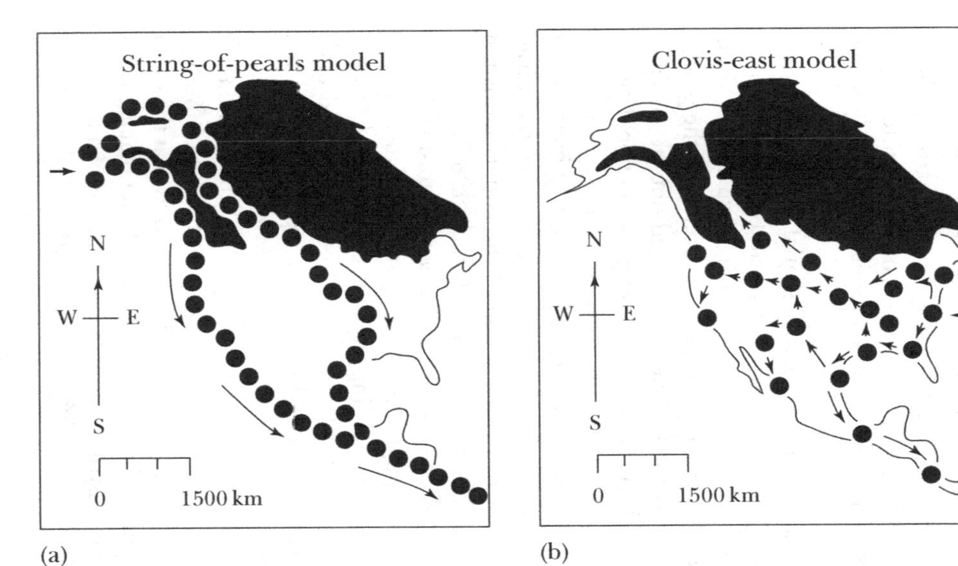

Figure 8.4 Spatially explicit models for the late-Pleistocene invasion of Paleoindians into North America, using least-cost pathway analysis. A. The "string-of-pearls" model, reflecting southward immigration and short dispersal jumps to staging areas along the Pacific Coast Corridor and the continental interior route through the Alberta Ice-free Corridor, then southeastward movement along major river valleys (Anderson and Gillam, 2000). B. Our "Clovis-east" model depicting invasion of Clovis people on to the central Atlantic Coast from an origin in the Iberian Peninsula of southwestern Europe via the pack-ice bridge across the North Atlantic Ocean. Dispersal to the North American interior is the reverse of directions shown in the "leap-frog" model (modified from Anderson and Gillam, 2000). Continental glaciers are shown in black for 14 000 BP. The staging areas are shown as 400-km-diameter circles.

the ecotonal focus of terrestrial, riparian, and maritime habitats. This ecological model for ecotone-constrained colonization proposes the stepwise invasion of the coastal-plain strand, then inland expansion along major stream drainages.

Anderson and Faught (1998) compiled a comprehensive database plotting the county locations where more than 12 000 Paleoindian projectiles have been found. For the conterminous United States, more than two-thirds of these fluted stone points are concentrated along the middle Atlantic Seaboard, the Southeast, southern New England, and the lower Midwest regions. Especially high concentrations of these points occur along the Atlantic Coast, indicating the prospect that even more Paleoindian sites may be found offshore on the continental shelf, sites inundated by postglacial sealevel rise. The projectile evidence supports the contention that Clovis people invaded the Atlantic Seaboard first, but not by way of the Pacific Northwest or Ice-free Corridor. Unfortunately, the timing for Early Paleoindian colonization is poorly documented in eastern North America (Morse *et al.*, 1996). The oldest accepted dates for fluted points in the unglaciated Southeast are 15 350 and 13 800 BP (Broster and Norton, 1996), both radiocarbon dates from the Johnson–Hawkins archaeological site near Nashville, Tennessee (Ellis *et al.*, 1998). These dates are older than the 13 500 BP dates reported for Clovis from the western and northeastern United States.

Using fluted "Classic Clovis" points as time markers for occupation, Anderson and Faught (1998) contended that Clovis people were not the first to establish a New World beachhead. They were, however, "wildly successful and archaeologically highly visible" as they leap-frogged from one "staging area" to the next (Anderson and Faught, 1998, p. 176). Clovis lithic requirements and hunting preferences focused their initial "search image" in exploring new regions. Their "high technology" pattern of group foraging selected for desired "resource patches" in particular kinds of landscapes (Kelly and Todd, 1988). Clovis invaders sought fresh water, large game animals, abundant and high-quality stone for tools, and a landscape setting that favored communication among increasingly dispersed, nomadic bands. As initial populations settled, grew and split off into new bands, riparian corridors facilitated social contact for renewing kinship bonds and seeking mates, and even shaped their seasonal round of food-gathering activities (Anderson and Faught, 1998).

As an alternative explanation, Clovis projectile points may not strictly reflect the rapid and geographically widespread colonization of Clovis populations. Rather, those stylistically distinctive and deadly hunting weapons may have reflected the cultural adoption of a "good idea," a technological solution increasing hunting effectiveness that readily spread among Paleoindian populations previously established across North and Central America (Adovasio and Pedler, 1997, p. 579; Surovell, 2000, p. 494).

We suggest a variation of the least-cost "leap-frog" model in order to account for the dense but widely scattered concentrations of Clovis points across southeastern North America (Figure 8.4). We propose a "Clovis-east" model in which the arrows for direction of migration would show a generally westward Clovis dispersal inland from the continental shelf of the western Atlantic Ocean. We contend that the archaeological evidence supports a founding population for Clovis people that arrived first along the Central Atlantic Seaboard by at least 15 600 BP. By 14 000 BP, these colonists dispersed along the Atlantic and Gulf coastal plains, then westward along major riverways into the continental interior (Figure 8.4). Whether or not this model is correct, a dilemma still remains. Where was the source of founding populations for Clovis people?

The Solutrean solution

Clovis people may have had a western European ancestry (Stanford and Bradley, 2000). Between 25 000 and 19 650 BP, a Late Paleolithic, Solutrean Culture occupied the coastal zone of Portugal, northern Spain and southwestern France. On the Iberian Peninsula, between 19 650 and 13 000 BP the Magdalenian Culture replaced the Solutrean Culture (Collins, 1998). The Solutreans and Magdalenians shared the style of "overshot flaking" of chert to fashion tools. Bounded northward by glaciers and polar desert until 15 600 BP, this southern sector of western Europe provided an ice-age haven for boreal and cool–temperate ecosystems (Huntley and Birks, 1983), as well as a major refuge for European people (Torroni *et al.*, 1998). Late-Pleistocene European people used bone, ivory, and stone to produce an elaborate assemblage of tools (like shaft wrenches for straightening atlatl darts), weapons (diagnostic spear points and large bifaces thinned or flaked along their base), and engraved limestone tablets (plaquettes).

The Old World stone-working technology of the Solutrean and Magdalenian craftsmen produced a sophisticated toolkit that is virtually identical to the New World assemblage of Clovis artifacts. For example, in North America, one of the largest collections of Clovis artifacts has been excavated at the Gault archaeological site in central Texas. This prehistoric chert quarry contained 30 engraved cobbles of limestone (Collins *et al.*, 1991), exhibiting the full "Solutrean/Magdalenian" stylistic range of superbly flaked Clovis artifacts (Collins, 1998).

Variants of Clovis projectiles from the southeastern United States have been described as Cumberland and Barnes styles that were often fluted on only one side, with a shallow basal flute (Morrow and Morrow, 1999). These projectile forms provide a plausible starting point for the stylistic drift of Clovis technologies in crafting projectiles that may have changed morphometrically into "fishtail points" knapped with shallow or no flutes as people migrated into South America (Morrow and Morrow (1999).

The fluted Clovis end blade may represent an aboriginal technical solution to improve success in hunting of marine mammals by kayak, with detachable harpoons cast by spear throwers (Dixon, 1999). Such harpoons would have made these first Americans well adapted to survival in an ice-floe and calving ice-shelf environment, whether in the North Pacific Ocean (Dixon, 2000) or the North Atlantic Ocean. The problematic time gap of several thousand years in the archaeological continuity between European source populations of Solutrean Culture and presumed Clovis descendents (Straus, 2000) may reflect a time interval during which migratory people lived offshore in marine ice-shelf environments.

The ice-age North Atlantic Ocean could have served as a viable Paleoindian route for the European peopling of eastern North America (Stanford and Bradley, 2000). These groups may have traveled on foot along the ice bridge of the marginal ice shelf, or they could have paddled skin boats among ice floes in the seasonally fluctuating floes of pack ice. Ice domes on land masses bordering the North Atlantic Ocean extended seaward as a persistent and laterally continuous ice shelf, providing over-ice access from northwestern Europe, to Iceland, Greenland, and northeastern North America from 24 000 until 15 600 BP (Ruddiman and McIntyre, 1981; Imbrie *et al.*, 1983; Delcourt and Delcourt, 1984; Ruddiman, 1987). Between 15 600 and 13 000 BP, the climatic boundary of the Polar Front Jet Stream deflected northward to as much as 60° N (Iceland), and warm, saline surface waters flowed as the North Atlantic Drift, substantially reducing seasonal sea-ice formation offshore of northwestern Europe (Ruddiman, 1987). North of the full-glacial position of the maritime polar front (Figure 8.2), winter pack ice formed south to the Solutrean haven of Iberia, spanned the North Atlantic at 48° N, and joined the American Delmarva Peninsula at about 40° N. With summertime-warming, ice floes formed as the pack-ice margin retreated back to the ice shelf as far as 60° N near Greenland.

We propose that the Solutrean route for New World colonization was tethered to the North Atlantic interface between glacial ice and ocean.

The ice-age juxtaposition of warm, nutrient-rich Gulf Stream waters, the accentuated climatic gradient imposed by the consistent, year-round position of the jet-stream location of the Polar Front, and the extreme seasonal fluctuation in extent of sea-ice would have produced a bountiful marine ecosystem with abundant food resources of fish, birds, marine seals (Family Phocidae), walrus (*Odobenus rosmarus*), whales (Order Cetacea), and other large mammals such as polar bears (*Ursus maritimus*). Today, such seasonal changes in high-latitude formation and spatial extent of sea-ice serve as the primary force driving marine biological productivity in the North Atlantic (Ferguson *et al.*, 2000). During the last glacial maximum, the "direct" 5000-kilometer route from the Iberian Peninsula to the Delmarva Peninsula was not necessarily across the open ocean (Figure 8.2). Rather, a plausible route for human migration was available in the seasonally shifting maze of open water leads among ice floes, forming sea-ice along the persistent ice-bridge crescent of glacial-ice shelf. Populations of nomads hunting marine mammals on the sea-ice would have had a replenishable supply of walrus ivory and bone available as raw material to make weapons. To replace stone blades, Solutrean travelers could even have quarried stone nodules, embedded in shear-band layers of "dirty ice" entrained within icebergs and rafted into the North Atlantic Ocean from continental glaciers.

Megapatch model

Using the estate settlers' model of Beaton (1991), Dixon (1999) offered a megapatch model for the geographic pattern and pathways for the first sustained colonization of the New World. Dixon proposed a Pacific Coastal Corridor as another contiguous land/sea route for Paleoindian exploitation of coastal resources. First, colonization would succeed on the west coast of the Americas, then colony offshoots would cross the fundamental megapatch or biome boundary of coastland lowlands to Cordilleran/Andean uplands. In this scheme, the Atlantic Coastal Plain of eastern North America might be invaded by Asian lineages crossing eastward over the narrow Panama Isthmus or traversing the southern tip of South America. Contemporaneous arrivals in the northeastern sector of North America via the Atlantic ice shelf were not ruled out (Dixon, 1999). Continental interiors and deglaciated northern Canada, however, would have been the last areas to have been settled. The megapatch model thus postulates demographic saturation of a culturally preferred biome before colonization efforts crossed ecotonal borders into new ecological habitats (Dixon, 1999).

Surovell (2000) developed an innovative model that provided a family or band unit-based mechanism for the effective colonization of biomes by Paleoindians. Recognizing that humans are not passive dispersal units but are instead family units or small bands with a long interval of child-rearing, this model incorporated a simple geometry of foraging behavior to demonstrate that the high mobility envisioned for Paleoindian hunter–gatherers is compatible with the high reproductive rates among females required to account for accelerated population growth and dispersal. By adopting a gender-specific land-use strategy that minimized workloads for mothers with young children, residential base camps could have been moved frequently to minimize foraging distances for women and children, and yet maximize the larger search radius necessary for success of males hunting larger game and gathering high-quality lithic materials for knapping tools.

Surovell's (2000) model is based on a lifeway strategy tailored to local habitats. It incorporates frequent movement of residential base camps, allowing hunter–foragers to access broad biomes while at the same time minimizing total distances traveled. In a homogeneous environment, costs of carrying children can be minimized by moving frequently over small increments of distance, with females required to forage less widely for plant resources than males hunting for game. Frequent, short moves not only permit short foraging distances around base camps but, over time, result in long-distance travel while minimizing daily walking distances, maintaining high reproductive rates, and resulting in high infant survival rates. In a patchy environment, longer-duration occupations would become optimal, however, because as resource patches become more widely spaced, the cost of residential mobility would increase while the cost of logistical mobility would remain constant (Surovell, 2000). The ultimate rate of human migration within megapatch biomes may therefore have been impeded by density-dependent ecological processes, conditioned by spatial and temporal variability in resource availability and by environmental heterogeneity (Dillehay, 1991).

KEYSTONE SPECIES DISPLACEMENT AND HUMAN HUNTING PRESSURE

Keystone species regulate ecosystem processes through predator–prey interactions (Colinvaux, 1978). Keystone species have been interpreted as top carnivore predators dampening the natural herd fluctuation and reducing prey populations. Alternatively, keystone taxa may be the predominant herbivore species accounting for structural modification of habitat by their grazing, browsing, or trampling along game trails, and maintaining a heterogeneous landscape patchwork experiencing plant-successional regrowth (McNaughton and Georgiadis, 1986; Senft *et al.*, 1987).

Diagnostic fluted projectile points from North America, including both Clovis and Folsom styles, date from 15 490 to as young as 6630 BP, with peak frequency represented by 49 dates tallied from 13 000 to 12 000 BP (Bonnischen *et al.*, 1987) (Figure 8.1). This overall mode of Paleoindian activity reflected an intensification of big-game hunting between 14 000 and 11 500 BP, as an ephemeral adaptive cultural response to massive climate change driven by accentuated seasonal contrast between winter and summer, to rapidly changing natural environments, and to declining megafaunal populations (Bonnischen *et al.*, 1987). The last records of late Paleoindian points may reflect persistence of communal hunting groups, perhaps following northward "the call of distant mammoths" and surviving as isolated, remnant populations scattered across deglaciated Canada (Ward, 1997).

Hunting pressures upon governing keystone species might have driven the megafaunal extinction process in three very different ways: (1) selective elimination of megaherbivores (the Overkill model of Martin), with subsequent population collapse of natural predators, scavengers, and commensal taxa ecologically dependent upon the plant-eaters (trophic cascades of extinction *sensu* Diamond, 1984); (2) initial human elimination of other competing keystone predators, unleashing herbivore boom–bust cycles as their populations are no longer "capped" by predation (the second-order Overkill model of Whitney-Smith, 1998); or (3) inadvertent transmission of hypercontagious disease or parasites by Paleoindians infecting New World mammalian faunas (the hyperdisease model of MacPhee and Marx, 1997). Testing these alternatives requires determining whether terminations of prey lead, lag, or are synchronized with their ecologically linked predators.

The youngest credible dates for predator extinctions in North America are intriguing, although not conclusive for this test. Reliable radiocarbon dates indicate relatively contemporaneous loss of megaherbivores with removal of predators, including sabertooth cat (*Smilodon*, 11 130 BP) and lion (*Panthera*, 12 170 BP). Less secure dates record even earlier terminations for cheetah (*Acinonyx*, 20 950 BP), the giant short-faced bear (*Arctodus*, 14 650 BP), and a relatively late time for the last scavenger dire wolf (*Canis dirus*, 11 500 BP) (Meltzer and Mead, 1985; Grayson, 1991).

Based upon ecological studies of contemporary African elephants (*Loxodonta africana*), rhinoceroses (*Diceros bicornis* and *Ceratotherium simum*), and elands (*Taurotragus oryx*), Owen-Smith (1987, 1989) interpreted the ecology of surviving "megaherbivores" as keystone species governing ecosystem dynamics of savannas and open woodlands. Annual mammal migrations connect a seasonal round of foraging and browsing along a predictable network of trails connecting permanent water sources. The daily herbivory demand for large quantities of edible, coarse fiber and the structural damage by pruning and trampling of woody shrubs and saplings serve as a disturbance regime of megaherbivores, suppressing tree growth and maintaining a landscape mosaic of open grasslands and early-successional forests. With late-Pleistocene removal of North American megaherbivores and concomitant climatic change, the heterogeneous and fine-grained patchwork of vegetation would succeed into the more homogeneous matrix belts of Holocene forests and prairie, leading in turn to the cascading loss of edge and early-successional habitats for disturbance-favored plants and small mammals. Hunting by Paleoindian bands would have substantially reduced and fragmented surviving populations of megaherbivores, leading not only to their extinction but also to extinction of predators, scavengers and commensals dependent upon them. In this way, *Homo sapiens* would have become the new keystone species regulating ecosystem dynamics in the late Pleistocene (Owen-Smith, 1989; Brown, 1995). Paradoxically, faunal extinction caused by human intervention could have been the cause, rather than the result, of the restructuring to a more homogeneous Holocene landscape because of elimination of a natural disturbance regime.

The greatest deterrent to Paleoindian entry into the New World may have been the giant short-faced bear (Geist, 1999). As a fearsome and effective hunter, short-faced bear posed both a potential predator upon humans and a competitor for the same animal food resources farther down through the food chain. Adapting the computer simulation model of Whittington and Dyke (1984), Whitney-Smith (1998) generated a second-order model for Overkill in which native people acted upon their perception of their environment, a behavioral "policy" by which humans preferentially hunted out the other predators that they feared. By intentionally reducing carnivore populations (by as little as 1.5 percent), simulation results indicate that the earliest Americans would have diminished the effectiveness of native predators in regulating sizes of megaherbivore populations. Before human populations could expand to occupy the role of keystone control, herbivores would have increased in density, outstripping their plant food resources and overshooting their carrying capacity. The herbivore boom would have provided an ephemeral opportunity of virtually endless mammal prey for Paleoindian hunters. The human population would increase, then diminish, tracking the boom–bust fluctuations of large prey animals. Structural simplification would be the consequence of widespread forest succession in eastern and western woodlands, while postglacial warmth, aridity, wildfire, and overgrazing by bison herds would perpetuate grassland prairies in the central belt of the continental interior.

Late-glacial human impact may have been less obvious in destabilizing predator–prey dynamics. Could the Paleoindian conquest of the New World have represented a form of prehistoric "ecological imperialism" (Crosby, 1986), inflicted by the diseases carried along by nomadic immigrants and their hunting packs of domesticated wolf-dogs? Desowitz (1997) noted that Amerindian immigrants, traversing the Arctic regions north of 60° N, passed through a "cold screen" that removed human-specific pathogens that formerly capped population vigor, fecundity, and survivorship. Removal of this population-limiting factor provided potential for a release effect in accelerating human population growth. This cold-sterilization freed Paleoindians from Old World intestinal parasites such as roundworm (*Ascaris lumbricoides*), whipworm (*Trichuris trichiura*), and hookworms (*Ancylostoma duodenale* and *Necator americanus*). These intestinal parasites require summer soil temperatures above 20 °C (68 °F) for worm eggs to hatch into infective stages of their life cycle. However, enhanced Milankovitch seasons of peak summer warmth (Figure 8.1) and warm soils may have allowed for limited pathogen survival through this Arctic "bottleneck."

The earliest Americans may, however, have constituted a vector for other kinds of disease transmission. Warm-blooded Paleoindians and their domesticated dogs (*Canis familiaris*) could have carried arboviruses, viruses capable of being transmitted through blood-sucking mosquitoes, ticks, and mites to a broad spectrum of hosts including mammals and birds. Desowitz (1997, p. 34) suggested that the trans-Arctic migrations of Paleoindians provided safe conduct of zoonitic arboviruses that "may have proved highly lethal to the immunologically naïve American wild mammals," perhaps contributing to megafaunal extinction. Effective viral transfer of infectious diseases among mammal species and genera would have radically altered the "immunological landscape" of the New World. Desowitz (1997, p. 73) speculated that "emerging pathogens [virulent diseases like Lassa, Ebola, Marburg, and HIV of AIDS] . . . are believed to have crossed the species barrier" particularly during times of habitat deterioration and climate change already stressing prospective populations of new animal hosts.

MacPhee and Marx (1997) analyzed ancient tissue of extinct megafauna, searching for paleo–viral evidence for such a hyperdisease. They hypothesize that virulent viruses were transmitted by Paleoindians with first contact, sparking a New World plague that crossed species boundaries, killed quickly, and reduced megafaunal populations over a wide area, prohibiting their population recovery. Such a rapid spread of infectious disease would have taken out large mammals from several ecological trophic classes, including herbivores, predators, and scavengers. Whether or not preClovis people occupied North America prior to 14 000 BP, a separate (second?) wave of late-Pleistocene human emigrés from 14 000 to 13 500 BP may have been the critical vector introducing catastrophic plague to the New World. Transmission of viral disease between alternate mammal hosts may have happened through blood transfer by mosquitos, mites, ticks, and chiggers, on the bloody stone tips of weapons humans used to wound prey, or from ingestion of human remains by predators.

As Paleoindian hunter–gatherers crossed Beringia, the Arctic filter stripped them of their Old World or "heirloom" species of parasites; the early Americans readily acquired a new suite of "souvenir" species of parasites and pathogens endemic to the New World (Dillehay, 1991). These new diseases may have been transmitted directly to humans, or indirectly through domesticated dogs, or by eating raw or poorly cooked meat of terrestrial and marine mammals, birds, fish, and shellfish. The reproductive success of colonizing humans in the Americas would have been impacted by suites of native pathogens and parasites they encountered, particularly as humans invaded temperate and tropical environments (Dillehay, 1991).

PALEOINDIAN POPULATION GROWTH

World-wide human population growth during the late Pleistocene was modeled by Butzer (1991; Figure 8.5). In the Upper Paleolithic, between 35 000 and 30 000 BP, the population attained a steady state at relatively low levels across Europe and Asia. With the development of the Solutrean Culture in western Europe by 25 000 BP and the micro-blade technology of eastern Asia by 17 000 BP, populations began to increase exponentially as humans exploited previously untapped resources using new technology. In North America, preClovis colonists may have been suppressed or eliminated by New World diseases. The Paleoindian period in North America marked a late-glacial acceleration in population growth at about 13 500 BP, reflecting

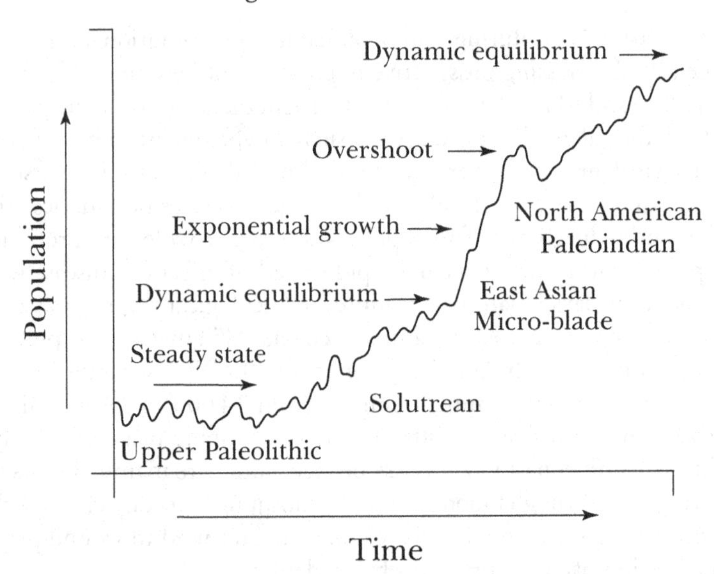

Figure 8.5 Model of late-Pleistocene and early-Holocene human population growth (adapted from Butzer, 1991).

(1) natural selection among preClovis groups for hardy, disease-resistant survivors; (2) relatively late entry of a Clovis founding group with greater immunity to disease; and/or (3) increased reproductive success of Clovis people with onset of late-glacial climatic amelioration and a cultural shift to exploit new food resources.

Paleoindians may have experienced a temporary over-expansion in populations, which was curbed after the late-Pleistocene extinction of megafauna (Butzer, 1991). There are several plausible explanations to account for such an "overshoot" phenomenon (Catton, 1982). First, Paleoindians may have decimated large mammals in a wave-front expansion, resulting in "overkill" of Pleistocene megafauna (Mosimann and Martin, 1975). Depletion of the megafaunal resource by over-specialized Paleoindian hunters would have made previously unrestricted rates of human population growth unsustainable, with a population adjustment occurring until new subsistence strategies were adopted by these hunter–foragers.

A second alternative is that there may have been two kinds of subsistence strategies operating at the same time during the Paleoindian phenomenon, with northern and western populations exploiting big game animals in tundra and plains environments that were relatively species-poor, and with northeastern and southeastern Paleoindians engaged in a more generalized foraging strategy within a biotic region characterized by more species-rich boreal and temperate forests (Meltzer and Smith, 1986). In this case, human exploitation of resources would have been optimized, both in the northern and the southern sectors of North America, leading to a rapidly expanding human population, until resource depletion curbed population growth at the time of megafaunal extinctions. This second scenario could have been achieved if the generalized eastern Clovis foragers were descendents of western Clovis people (Morrow and Morrow, 1999; Dixon, 1999) or if they were derived from Solutrean immigrants (Stanford and Bradley, 2000). The geographic implications are that the ecologically more diversified southern and southeastern Clovis cultures (Anderson and Gillam, 2000; Tankersley, 1998) expanded northward with emerging early-Holocene forest ecosystems, their populations displacing and replacing their distant cousins, the alpine–arctic specialists adapted to relatively open grasslands and parklands. As the specialist Clovis hunters tracked northward following retreating herds of caribou and proboscideans, their cultural adaptive lifeways could no longer be maintained along the attenuated periglacial zone flanking the disintegrating Laurentide Ice Sheet. Remaining Clovis descendents would trace their "southern heritage" and colonizing success to a broadly based hunting–foraging strategy.

CONCLUSIONS: *HOMO SAPIENS* AS A KEYSTONE SPECIES

Major climate change at the late Pleistocene/Holocene transition profoundly altered regional climate patterns in temperature and precipitation regimes, and drove radical changes in physical environments and restructuring of plant and animal communities. These environmental changes not only contributed to megafaunal extinction and wholesale geographic shifts in the ranges of small mammal species, they also amplified the ephemeral cultural success of nomadic people's strategy of hunting large mammals.

Particularly from 17 000 to 11 500 BP, Milankovitch cycles enhanced seasonal contrast in climate and served as a forcing function for environmental change as glaciers melted and available land area on the North American continent nearly doubled. During the late-glacial interval, vast deglaciated landscapes were available for colonization by "pulse" pioneer ecosystems of arctic and boreal plant and animal species (Birks, 1986). The pulse of nutrients released by melting glacial ice, wind-deposited silts, and from weathering of newly exposed tills and outwash substrate led to increased biological productivity, temporarily increasing the carrying capacity of megafaunal populations. This ephemeral peak in population numbers for prey animals coincided with their behavioral "tethering" to seasonally restricted water holes during late-glacial times of very cold winters and very hot summers. The boom-then-bust of megafaunal populations reflected an initial positive response to expanded distribution range, followed by population collapse from the multiple stresses including (1) changing physical habitats used for shelter; (2) dietary shifts imposed by changing plant resources; (3) reproductive dysfunction; (4) seasonal crises of severe water shortage; and (5) density dependence problems tied to overcrowding, disease transmission, and overgrazing.

PreClovis bands from widely disparate areas may have responded to this situation as a new resource opportunity. Cultural perception of the increase in numbers of megaherbivores and their seasonal aggregation and vulnerability resulted in creating or adopting the technologic innovations of new Clovis-style, fluted stone weapons. The combination of adopting a lifeway of big-game hunting coupled with rapid growth of Paleoindian populations set the North American stage for overkill as prey populations were excessively harvested and depleted.

The megafaunal extinctions, largely finished by 11 500 BP, marked the end of the Paleoindian lifeway of big-game hunting. Humans were forced to seek new solutions in how they sought food, where they settled, and how they crafted tools. The early-Holocene transition phase of this macroscale, glacial–interglacial adaptive cycle generated wholesale panarchical restructuring of ecosystems and set the stage for the regional emergence of human-managed ecosystems as new adaptive panarchical levels (Figure 3.5).

Ecological models for human colonization of North America contrast the possible roles of physical terrains, ecotones, and biome megapatches in shaping the regional patterns of Paleoindian invasion. The geographic routes taken by founding populations of nomads, the timing of their entries, and their lifeway strategies as either transient explorers or estate settlers, all were important factors in the role of *Homo sapiens* as a keystone species in contributing to the late-Pleistocene panarchical restructuring of regional ecosystems 11 500 calendar years ago.

PART III

Application and synthesis

OVERVIEW

In Chapter 9 we summarize the implications of viewing Holocene human ecosystems as adaptive and panarchical in nature. In the United States, land management plans have been predicated on the premise that prehistoric humans had no lasting influence on native ecosystems. Well-documented paleoecological and archaeological studies clearly indicate that this assumption is not warranted. Since their arrival on the North American continent, Native Americans were part of an evolving set of human ecosystems. Through time, increasing complexity of the human–ecosystem panarchy arose from both cultural and ecological interactions. These interactions were effective at all levels of biological organization, from genetic, to population, community, landscape, and regional levels (Figure 3.4). No single arbitrary time line is appropriate as a focus for land management, since ecosystems have been evolving continuously in response to late-Quaternary environmental change and human activities. Scientific studies combining paleoecological and archaeological techniques provide a context for the trajectory of change on a local, regional, and continent-wide basis. Understanding this context is essential for guiding future efforts to manage biological diversity.

9

The ecological legacy of prehistoric Native Americans

THE MYTH OF THE PRISTINE FOREST

The explicitly stated choice of targets for ecological land management efforts in the United States mandates a "return to the presettlement equilibrium," defined as a preColumbian baseline (Peacock, 1998; Kimmerer, 2000; Kloor, 2000). Since the inception of academic ecology in the United States at the turn of the twentieth century, scientific investigations have been predicated on the assumption that the North American continent has been relatively undisturbed by human activities, giving investigators a series of "natural experiments" from which fundamental relationships of plants and animals to their environment can be ascertained (Wright, 1977; McIntosh, 1985; Risser *et al.*, 1984). During much of the twentieth century, ecologists also viewed the original vegetation and fauna of North America as being in a steady state, in equilibrium with climate and other environmental factors (McIntosh, 1985; Golley, 1984).

Both the assumption that ecosystems existing prior to the time of European American settlement were untouched by Native American activities and that they were in a state of equilibrium have been questioned, based upon a number of lines of reasoning, as well as historical and prehistoric information (Denevan, 1992; Kay, 1995; Hamel and Buckner, 1998; Peacock, 1998; Krech, 1999; Redman, 1999; Kimmerer, 2000; Minnis and Elisens, 2000).

"The forests the settlers saw"

Ecological methods have been developed to assess the distribution and composition of vegetation existing at the time of the westward expansion of European American settlement, using records from the General Land Office Surveys, under the assumption that "the forests the settlers saw" were representative of pristine "presettlement" wilderness (Bourdo, 1956, 1983;

Marschner, 1959, 1974; Williams, 1989). The widespread notion that the vegetation the pioneers saw represented the "potential natural vegetation," that is, the vegetation that would regenerate in the absence of human influence (Küchler, 1964), became the basis for guidelines to conserve representative remnants of natural vegetation as well as for efforts to restore altered vegetation to its natural state (Kimmerer, 2000; Kloor, 2000).

Impacts of European American settlement

Based upon ethnographic accounts and reconstructed population estimates for the year AD 1492, Denevan (1992) hypothesized that the landscape of North America at the time of Columbus was an anthropogenic one rather than a natural one, having been transformed over millennia by Native Americans through land clearance, farming, and their use of fire. Denevan (1992) estimated that there were more than 50 million indigenous people in the New World in preColumbian times, with nearly 4 million in North America, and the remainder concentrated in Mexico (17.2 million), Central America (5.6 million), the Caribbean (3.0 million), the Andes (15.7 million) and lowland South America (8.6 million). The overall population densities may have been as high as those of much of fifteenth-century Europe.

With initial European contact and spread of infectious diseases, indigenous human populations declined by up to 90 percent throughout the New World (Dobyns, 1983; Crosby, 1986; Ramenofsky, 1987; Hudson, 1997; Cook, 1998; Henige, 1998). With decreased hunting pressure, herds of bison increased on the Great Plains. Abandoned "tallahassees" of Indian old-fields underwent secondary succession to regenerated forests over much of eastern North America. Native plants once used by Native Americans for construction materials and weaving, for example American cane (*Arundinaria gigantea*), expanded to form successional plant communities on abandoned Indian old-fields throughout the floodplains of major rivers of the Southeast (Platt and Brantley, 1997; Hamel and Buckner, 1998).

European American colonization of the Eastern Seaboard occurred slowly, with a long initial phase of homesteading and land clearance concentrated along the coastline and the lower Mississippi River Valley from 350 to 150 BP (Williams, 1989). With the development of the General Land Office Surveys and railroad systems, by 150 BP much of the interior of the North American continent became accessible for rapid exploitation of both game (e.g., bison and passenger pigeons) and timber (e.g., the great white pine forests of the Upper Great Lakes region) (Williams, 1989; Whitney, 1994). But even by 200 BP, the landscapes that were previously utilized extensively by Native Americans had been abandoned and had largely regenerated as late-successional ecosystems. By the time that most Europeans saw them, the landscapes of eastern North America appeared to be a wilderness (Hamel and Buckner, 1998; Denevan, 1992). Rather than the forest primeval, however, the vegetation was composed of secondary stands that had regrown during just one successional cycle after Native American abandonment (Byrne and Finlayson, 1998; Hamel and Buckner, 1998).

Setting priorities for land management

Redman (1999) suggested two considerations for environmental decision-making, based upon his extensive analysis of the prehistoric record of human interactions with the environment. He stated, "There is no absolute when one refers to the natural state of the environment . . . Most environments, as we now see them, have developed under pressures created by the presence of humans . . . [these] transformations range from obvious to subtle . . . A definition of an ideal, or best, environment is conditioned by human values and objectives" (Redman, 1999, pp. 202–3).

Ecological land management is predicated on selecting an appropriate landscape state to maintain, one that may be selected for its aesthetics, biodiversity, productivity, or recreational value (Redman, 1999). If it is appropriate to manage a landscape as it was prior to European American settlement, then it is necessary to document both the patterns and dynamic processes shaping the landscape at that time.

Paleoecological and archaeological studies reinforce the conclusion that throughout the late Pleistocene and Holocene, environmental change has been the rule rather than the exception. Climate changes on all scales in space and time, and changes in the distribution and composition of both flora and fauna occur continually in response to changes in climate. Both natural and anthropogenic disturbances have been important influences on ecosystem development through the Holocene. Therefore, determining the "best" course for ecological land management related to any particular timeline is an arbitrary and subjective decision (Peacock, 1998; Hamel and Buckner, 1998; Redman, 1999; Kimmerer, 2000).

INTEGRATING VIEWPOINTS: THE ROLE OF TRADITIONAL ECOLOGICAL KNOWLEDGE

Because of increased understanding of the importance of human influences on the biota, academic ecologists and Native Americans who share a common interest in conservation of biological diversity and ecologically/historically correct restoration and management of native ecosystems have combined efforts. Increasingly, traditional ecological knowledge has come to the forefront as a valid subfield of applied ecology (Blackburn and Anderson, 1993; Berkes, 1999; Ford and Martinez, 2000). As Morse *et al.* (1996, p. 334) have stated succinctly, "No one lives for the distant future, but there is a tradition of the past that guides our futures."

Insights from traditional ecological knowledge

Berkes (1999, p. 8) defined traditional ecological knowledge as "a cumulative body of knowledge, practice, and belief, evolving by adaptive processes and handed down through generations by cultural transmission about the relationship of living beings (including humans) with one another and with their environment." As such, traditional ecological knowledge is a "knowledge-practice-belief complex" (Figure 3.1), an attribute of indigenous societies that, historically, have continuity in particular practices of resource use (Berkes *et al.*, 2000). In that traditional ecological knowledge has been gathered over the course of many generations by people whose lives depended upon the information gained, it tends to be both incremental in accumulation and holistic in outlook (Berkes *et al.*, 2000).

Central to traditional ecological knowledge is a sense of place, wherein people live within nature, not apart from nature. Traditional ecological knowledge thus differs from Western philosophy in considering that humans are part of nature, not autonomous from and in control of the natural world. The Native American worldview is spatially oriented, in contrast to western European thought, which is temporal in perspective. A basic concept embodied within traditional ecological knowledge that is similar to scientific ecological thinking is that all things are connected – that is, "that no single organism can exist without the web of other life forms that surround it and make its existence possible" (Pierotti and Wildcat, 2000). This recognition of interrelationships between different species is a concept closely related to community ecology, although Native people view themselves as part of that community whereas academic ecologists typically study intraspecific and interspecific interactions of plants and animals in isolation from their own role as observers (Pierotti and Wildcat, 2000).

Native American reservations occupy more land in the United States than do national parks and national forests, and therefore contain a significant percentage of the native fauna and flora of North America north of Mexico (Nabhan, 2000a, b). Conservation of this biodiversity depends in part on enlisting the efforts of "para-ecologists" who are well-versed in practices based on traditional ecological knowledge (Nabhan, 2000a, b). The cultural management practices of Native Americans provide an essential complement to scientific ecological knowledge that may be critical to conservation and restoration efforts (Ford and Martinez, 2000). The focus of academic conservationists has largely centered on preserving species, rather than on understanding the processes and interactions that are essential to their survival (Franklin, 1993; Pickett *et al.*, 1992; Delcourt and Delcourt, 1998b). Indigenous people's knowledge of ecological interactions among humans, other animals, plants, and their habitats are often embedded within a tribe's linguistic tradition, which itself is in danger of being lost because of acculturation and disuse of native languages during the twentieth century. Native words for a number of ecological processes, ranging from predator–prey interactions to microhabitat relationships, form the basis of hypotheses to be tested using scientific ecological methods, a field of inquiry termed "ethnoecology" by Nabhan (1989, 2000a, b).

Peacock and Turner (2000) and Turner *et al.* (2000) described the relationship of foraging practices of indigenous people of British Columbia to maintaining biological diversity at species, community, and landscape levels. In much the same way that Native Americans in California used tillage as a form of intermediate disturbance to sustain the vigor and diversity of species of herbs with underground rhizomes, tubers, and corms (Anderson and Nabhan, 1991; Anderson and Moratto, 1996; Anderson, 1997), people of the Northwest Coast developed methods for sustainable harvests of native plants such as balsamroot (*Balsamorhiza sagittata*) and yellow avalanche lily (*Erythronium grandiflorum*). Archaeological evidence for tillage and consumption of these plants, evidenced by archaeological remains of handles of root-digging sticks and cooking pits, date to as long ago as 2400 BP. Historically, populations of these plant species have been observed to decline, both because of fire suppression and because of introduction of domestic cattle and non-native species of plants (Turner *et al.*, 2000).

Limitations of traditional ecological knowledge

Indigenous people have managed biological diversity to obtain a flow of natural resources and ecological services on which they depend for their livelihood. This worldview is consistent with the assumption that nature is not in a steady state, but rather it is in constant change, and that societies must respond by continually adjusting and evolving, termed "adaptive management" by Holling (1978) and Gunderson *et al.* (1995). Berkes *et al.* (2000) cautioned, however, that although many traditional practice and belief systems have resulted in effective management of native plant and animal resources, some societies became maladapted through time because of changing conditions that were beyond their experience base. Therefore, while rediscovery of traditional systems of knowledge and resource management offers a number of possibilities for sustainability of both biological species and ecological interactions, not all traditional practice is ecologically wise, and ecological land management is to indigenous people a fundamentally different concept than that of Western ecologists (Berkes *et al.*, 2000). Thus the practices of indigenous people have, in some cases, made their societies vulnerable to collapse during times of rapid environmental change during which they exceeded a "variation threshold for extinction" (Fagan *et al.*, 1999), as was the case for Cahokia in the Late Mississippian cultural period.

Traditional ecological knowledge may be most effective as a complementary source of ecological information for those geographic regions in which a cultural heritage has remained intact over the past several thousand years, such as in the Pacific Northwest. In other regions, for example in the Phoenix Basin of the American Southwest, the lifeways of historic Native Americans such as the Pima differ greatly from those of their ancient Hohokam predecessors. The Pima immigrated into central Arizona about 400 BP from coastal Mexico (Rea, 1997). They adopted many of the farming methods introduced by Europeans, including growing of winter wheat and raising of cattle, and even the type of commercially grown cotton today called "Pima" is a different species than that grown prehistorically (Rea, 1997). Extrapolation of Pima knowledge of plant–animal–environment interactions to a time before European contact and conquest thus cannot yield a complete understanding of the natural, human-influenced ecological system that existed before introduction of exotic species from the Old World.

CONCLUSIONS: PREHISTORIC NATIVE AMERICANS AS AGENTS OF ECOLOGICAL CHANGE

From the empirical evidence available from the paleoecological and archaeological records, particularly that focused on eastern North America, we draw a series of five fundamental conclusions concerning prehistoric Native Americans as agents of ecological change.

(1) Through the past 15 000 years, since the close of the last Ice Age, humans have been part of the North American ecological setting. Overall, the effects of Native Americans have been cumulative over time. The degree to which we can detect their influence depends in part upon the population size of the prehistoric population, which to be interpretable must reach a level at which the material culture they left behind is archaeologically visible. To be effective in studying the ecological roles of prehistoric humans, both the archaeological and the paleoecological techniques used must be sensitive to site formation and preservation as well as to the spatial and temporal scales at which humans interact with other organisms and with their abiotic environment. Specific, interdisciplinary case studies are necessary to understand the trajectory of change at a given place over time as a context for determining the relative effects of environmental changes, disturbance regimes, and human influences in shaping landscapes and determining biological diversity.

(2) During the present interglacial, across North America there was a panarchy of human ecosystems distributed along gradients in space and through time. Human cultures have evolved as part of nature, not apart from it, but humans are a unique ecological factor in having both cultural interactions and accumulated knowledge based on past generations of experience. Prehistoric Native American cultures were adaptive. Aboriginal societies were sustainable within bounds determined by the relative degree of climatic stability, proximity to geomorphic and habitat fragmentation thresholds, and major cognitive, cultural, and population thresholds such as those reached at the inception of native plant domestication by 4500 BP and with the shift to intensive agriculture by 1000 BP.

(3) Prehistoric human activities had long-term consequences. At the end of the Pleistocene, nomadic human hunters were one of many factors in the extinction of Pleistocene mammals. In the mid- and late Holocene, hunting and foraging strategies, as well as plant domestication activities, enhanced biological diversity at local to regional spatial scales and at population, species, community, and landscape levels. With increasingly unstable late Holocene climates, prehistoric Native Americans also contributed to environmental degradation by over-exploitation of wood resources and intensive cultivation of non-indigenous crops.

(4) There is no such thing as a Holocene, "presettlement," natural environment untouched by human hands in North America. The conception that Native Americans existed either apart from or in harmony with nature is incorrect, and based on false impressions and stereotypes first

developed by European American colonists. Previous comparisons of the Holocene history of Europe and North America as human-impacted landscapes versus pristine wilderness were exaggerated and are no longer a useful basis for either long-term ecological studies or for ecological land management efforts.

(5) Studies of historical documents and traditional ecological knowledge transmitted through oral histories to the current generation of Native Americans give a partial and subjective glimpse into the past. Ethnographic accounts and linguistic analysis therefore provide information that is complementary to, but that is not a substitute for, objective data from archaeology and Quaternary paleoecology. Rigorous analysis of archaeological evidence for site formation, including ethnobotanical and zooarchaeological studies, placed in an absolute time frame, is, however, essential for correct interpretation of prehistoric lifeways and changes in human populations through time. Paleoecological data, including fossil pollen, plant macrofossils, and charcoal particles, collected from wetland sites in close proximity to well-studied archaeological sites, make it possible to reconstruct the direct influences of prehistoric humans on their ecological setting. Geomorphic evidence of changes in river regimes and soil development, as well as tree-ring analysis of climate changes, are crucial adjuncts to developing an integrated view of prehistoric human adaptations to paleoenvironments. Such syntheses of paleoenvironmental and cultural information are central to deciphering the particular Holocene trajectory of any given landscape. Such analyses are crucial to obtaining a more complete understanding of the role of prehistoric Native Americans as agents of ecological change.

References

Abel, T. (1998). Complex adaptive systems, evolutionism, and ecology within anthropology: interdisciplinary research for understanding cultural and ecological dynamics. *Georgia Journal of Ecological Anthropology*, **2**, 6–29.

Adams, K. R. (1994). A regional synthesis of *Zea mays* in the prehistoric American Southwest. In *Corn and Culture in the Prehistoric New World*, ed. S. Johannessen and C. A. Hastorf. Boulder, CO: Westview Press, pp. 273–302.

Adovasio, J. M., and Page, J. (2002). *The First Americans, in Pursuit of Archaeology's Greatest Mystery*. New York, NY: Random House, Inc.

Adovasio, J. M., and Pedler, D. R. (1997). Monte Verde and the antiquity of humankind in the Americas. *Antiquity*, **71**, 573–80.

Albert, L. E., and Wyckoff, D. G. (1981). Ferndale Bog and Natural Lake: five thousand years of environmental change in southeastern Oklahoma. *Oklahoma Archaeological Survey, Studies in Oklahoma's Past*, 7, 1–125.

Allen, T. F. H., and Starr, T. B. (1982). *Hierarchy: Perspectives for Ecological Complexity*. Chicago, IL: University of Chicago Press.

Anderson, D. G. (1994). *The Savannah River Chiefdoms: Political Change in the Late Prehistoric Southeast*. Tuscaloosa, AL: University of Alabama Press.

(1995a). Recent advances in Paleoindian and Archaic period research in the southeastern United States. *Archaeology of Eastern North America*, **23**, 145–76.

(1995b). Paleoindian interaction networks in the Eastern Woodlands. In *Native American Interactions: Multiscalar Analyses and Interpretations in the Eastern Woodlands*, ed. M. S. Nassaney and K. E. Sassaman. Knoxville, TN: University of Tennessee Press, pp. 3–26.

(1996). Models of Paleoindian and Early Archaic settlement in the Lower Southeast. In *The Paleoindian and Early Archaic Southeast*, ed. D. G. Anderson and K. E. Sassaman, Tuscaloosa, AL: University of Alabama Press, pp. 29–57.

(2001). Climate and culture change in prehistoric and early historic eastern North America. *Archaeology of Eastern North America*, **29**, 143–86.

Anderson, D. G., and Faught, M. K. (1998). The distribution of fluted Paleoindian projectile points: update 1998. *Archaeology of Eastern North America*, **26**, 163–87. The spatial database for fluted Paleoindian points can be accessed online at http://www.cr.nps.gov/seac/paleoind.htm. The county-level documentation for Paleoindian localities can be viewed online at http://www.cast.uark.edu/local/catalog/national/html/Archaeology.htmldir/USflutedens.html/.

Anderson, D. G., and Gillam, J. C. (2000). Paleoindian colonization of the Americas: implications from an examination of physiography, demography, and artifact distribution. *American Antiquity*, **65**, 43–66.

(2001). Paleoindian interaction and mating networks: reply to Moore and Moseley. *American Antiquity*, **66**, 530–35.

Anderson, D. G., and Hanson, G. T. (1988). Early Archaic settlement in the southeastern United States: a case study from the Savannah River valley. *American Antiquity*, **53**, 262–86.

Anderson, D. G., and Sassaman, K. E. (1996). Modeling Paleoindian and Early Archaic settlement in the Southeast: a historical perspective. In *The Paleoindian and Early Archaic Southeast*, ed. D. G. Anderson and K. E. Sassaman. Tuscaloosa, AL: University of Alabama Press, pp. 16–28.

Anderson, E. (1952). *Plants, Man, and Life*. Berkeley, CA: University of California Press.

Anderson, M. K. (1997). From tillage to table: the indigenous cultivation of geophytes for food in California. *Journal of Ethnobiology*, **17**, 149–69.

Anderson, M. K., and Moratto, M. J. (1996). Native American land-use practices and ecological impacts. In *Sierra Nevada Ecosystem Project: Final Report to Congress, Volume 2: Assessments and Scientific Basis for Management Options*. Davis, CA: University of California, Centers for Water and Wildland Resources, pp. 187–206.

Anderson, M. K., and Nabhan, G. P. (1991). Gardeners in Eden. *Wilderness*, **55**, 27–30.

Andrews, R. L., and Adovasio, J. M. (1996). The origins of fiber perishables production east of the Rockies. In *A Most Indispensable Art, Native Fiber Industries from Eastern North America*, ed. J. B. Peterson. Knoxville, TN: University of Tennessee Press, pp. 30–49.

Asch, D. L. (1994). Aboriginal specialty-plant cultivation in eastern North America: Illinois prehistory and a postcontact perspective. In *Agricultural Origins and Development in the Midcontinent, Reports No. 19*, ed. W. Green. Iowa City, IA: Office of the State Archaeologist, University of Iowa, pp. 25–86.

Asch, N. B., and Asch, D. L. (1978). The economic potential of *Iva annua* and its prehistoric importance in the Lower Illinois Valley. In *The Nature and Status of Ethnobotany, Anthropological Papers 67*, ed. R. Ford. Ann Arbor, MI: University of Michigan Museum of Anthropology, pp. 300–41.

Baden, W. W. (1987). A dynamic model of stability and change in Mississippian agricultural systems. Ph.D. dissertation. Knoxville, TN: Department of Anthropology, University of Tennessee.

Bareis, C. J., and Porter, J. W. (eds.). (1984). *American Bottom Archaeology: A Summary of the FAI-270 Project Contribution to the Culture History of the Mississippi River Valley*. Urbana, IL: University of Illinois Press.

Barnosky, C. W., Anderson, P. M., and Bartlein, P. J. (1987). The northwestern U.S. during deglaciation; vegetational history and paleoclimatic implications. In *North America and Adjacent Oceans during the Last Deglaciation, The Decade of North American Geology (DNAG) Vol. K-3*, ed. W. F. Ruddiiman and H. E. Wright, Jr. Boulder, CO: Geological Society of America, pp. 289–321.

Barrie, J. V., and Conway, K. M. (1999). Late Quaternary glaciation and postglacial stratigraphy of the Northern Pacific margin of Canada. *Quaternary Research*, **51**, 113–23.

Beaton, J. M. (1991). Colonizing continents: some problems from Australia and the Americas. In *The First Americans: Search and Research*, ed. T. D. Dillehay and D. J. Meltzer. Boca Raton, FL: CRC Press, pp. 209–30.

Beck, M. W. (1996). On discerning the cause of late Pleistocene megafaunal extinctions. *Paleobiology*, **22**, 91–103.

Bennett, K. D. (1997). *Evolution and Ecology: The Pace of Life*. Cambridge, UK: Cambridge Studies in Ecology, Cambridge University Press.

Berkes, F. (1999). *Sacred Ecology: Traditional Ecological Knowledge and Resource Management*. Philadelphia, PA: Taylor and Francis.

Berkes, F., Colding, J., and Folke, C. (2000). Rediscovery of traditional ecological knowledge as adaptive management. *Ecological Applications*, **10**, 1251–62.

Binford, L. R. (1980). Willow smoke and dogs' tails: hunter-gatherer settlement systems and archaeological site formation. *American Antiquity*, **45**, 4–20.

Binford, M. W., Brenner, M., Whitmore, T. J., Higuera-Gundy, A., Deevey, E. S., and Leyden, B. (1987). Ecosystems, paleoecology, and human disturbance in subtropical and tropical America. *Quaternary Science Reviews*, **6**, 115–28.

Birks, H. H., Birks, H. J. B., Kaland, P. E., and Moe, D. (1988). *The Cultural Landscape – Past, Present, and Future*. Cambridge, UK: Cambridge University Press.

Birks, H. J. B. (1986). Late-Quaternary biotic changes in terrestrial and lacustrine environments, with particular reference to north-west Europe. In *Handbook of Holocene Palaeoecology and Palaeohydrology*, ed. B. E. Berglund. New York, NY: John Wiley and Sons, pp. 3–65.

(1988). Conclusions. In *The Cultural Landscape – Past, Present, and Future*, ed. H. H. Birks, H. J. B. Birks, P. E. Kaland, and D. Moe. Cambridge, UK: Cambridge University Press, pp. 463–6.

Black, B. A., and M. D. Abrams. (2001). Influences of Native Americans and surveyor biases on metes and bounds witness-tree distribution. *Ecology* **82**, 2574–86.

Blackburn, T., and Anderson, K. (1993). Introduction: managing the domesticated environment. In *Before the Wilderness: Environmental Management by Native Californians*, ed. T. Blackburn and K. Anderson. Menlo Park, CA: Ballena Press, pp. 15–25.

Bloom, A. L. (1971). Glacial-eustatic and isostatic controls of sea level since the last glaciation. In *Late Cenozoic Glacial Ages*, ed. K. K. Turekian. New Haven, CT: Yale University Press, pp. 355–90.

Bonnischen, R., and Turnmire, K. L. (eds.). (1999a). *Ice Age People of North America: Environments, Origins, and Adaptations.* Corvallis, OR: Oregon State University Press and the Center for the Study of the First Americans.

(1999b). An introduction to the peopling of the Americas. In *Ice Age People of North America: Environments, Origins, and Adaptations*, ed. R. Bonnischen and K. L. Turnmire. Corvallis, OR: Oregon State University Press and the Center for the Study of the First Americans, pp. 1–26.

Bonnischen, R., Stanford, D., and Fastook, J. L. (1987). Environmental change and developmental history of human adaptive patterns; the Paleoindian case. In *North America and Adjacent Oceans during the Last Deglaciation, the Geology of North America, DNAG Vol. K-3*, ed. W. F. Ruddiman and H. E. Wright, Jr. Boulder, CO: Geological Society of America, pp. 403–24.

Bormann, F. H., and Likens, G. E. (1979). *Pattern and Process in a Forested Ecosystem.* New York, NY: Springer-Verlag.

Bourdo, E. A., Jr. (1956). A review of the General Land Office Survey and of its use in quantitative studies of former forests. *Ecology*, **37**, 754–68.

Bourdo, E. A., Jr. (1983). The forest the settlers saw. In *The Great Lakes Forest: An Environmental and Social History*, ed. S. L. Flader. Minneapolis, MN: University of Minnesota Press, pp. 3–16.

Braun, E. L. (1950). *Deciduous Forests of Eastern North America.* Philadelphia, PA: Blakiston.

Broster, J. B., and Norton, M. R. (1996). Recent Paleoindian research in Tennessee. In *The Paleoindian and Early Archaic Southeast*, ed. D. G. Anderson and K. E. Sassaman. Tuscaloosa, AL: University of Alabama Press, pp. 288–97.

Brown, J. H. (1995). *Macroecology.* Chicago, IL: University of Chicago Press.

Brown, M. D., Hosseini, S. H., Torroni, A., *et al.* (1998). mtDNA Haplogroup X: an ancient link between Europe/Western Asia and North America? *American Journal of Human Genetics*, **63**, 1852–61.

Bryant, V. M., Jr., and Holloway, R. G. (1985). *Pollen Records of Late-Quaternary North American Sediments.* Dallas, TX: American Association of Stratigraphic Palynologists Foundation.

Butzer, K. W. (1975). The ecological approach to archaeology: are we really trying? *American Antiquity* **40**, **1**, 106–11.

Butzer, K. W. (1982). *Archaeology as Human Ecology.* Cambridge, UK: Cambridge University Press.

Butzer, K. W. (1991). An Old World Perspective on potential Mid-Wisconsinan settlement of the Americas. In *The First Americans: Search and Research*, ed. T. D. Dillehay and D. J. Meltzer. Boca Raton, FL: CRC Press, pp. 137–56.

Byrne, R., and McAndrews, J. H. (1975). Pre-Columbian purslane (*Portulaca oleracea* L.) in the New World. *Nature*, **253**, 726–7.

Byrne, R., and Finlayson, W. D. (1998). Iroquoian agriculture and forest clearance at Crawford Lake, Ontario. In *Iroquoian Peoples of the Land of Rocks and Water, A.D. 1000–1650: A Study in Settlement Archaeology*, ed. W. D. Finlayson, Vol. 1. London, Ontario: London Museum of Archaeology, University of Western Ontario, pp. 94–107.

Campbell, C., and Campbell, I. D. (1992). Pre-Contact settlement pattern in southern Ontario: simulation model for maize-based village horticulture. *Ontario Archaeology*, **53**, 3–25.

Campbell, I. D., and Campbell, C. (1994). The impact of Late Woodland land use on the forest landscape of southern Ontario. *Great Lakes Geographer*, **1**, 21–9.

Campbell, I. D., and McAndrews, J. H. (1993). Forest disequilibrium caused by rapid Little Ice Age cooling. *Nature*, **366**, 336–8.

(1995). Charcoal evidence for Indian-set sites: a comment on Clark and Royall. *The Holocene*, **5**, 369–79.

Carmean, K., and Sharp, W. E. (1998). Not quite Newt Kash: three small rock-shelters in Laurel County. In *Current Archaeological Research in Kentucky*, ed. C. D. Hockensmith, K. C. Carstens, C. Stout, and S. J. Rivers, Vol. 5. Frankfort, KY: Kentucky Heritage Council, pp. 47–58.

Catton, W. R., Jr. (1982). *Overshoot: The Ecological Basis of Revolutionary Change*. Urbana, IL: University of Illinois Press.

Chapman, C. H. (1975). *The Archaeology of Missouri*, vol. 1. Columbia, MO: University of Missouri Press.

Chapman, J. (1975). *The Rose Island Site and the Bifurcate Tradition*. Reports of Investigations 14. Knoxville, TN: University of Tennessee, Department of Anthropology.

Chapman, J. (1994). *Tellico Archaeology: 12,000 Years of Native American History* (revised edition). Knoxville, TN: University of Tennessee Press.

Chapman, J., and Adovasio, J. M. (1977). Textile and basketry impressions from Icehouse Bottom, Tennessee. *American Antiquity*, **42**, 620–5.

Chapman, J., and Crites, G. D. (1987). Evidence for early maize (*Zea mays*) from the Icehouse Bottom site, Tennessee. *American Antiquity*, **52**, 352–4.

Chapman, J., and Shea, A. B. (1981). The archaeobotanical record: Early Archaic period to Contact in the lower Little Tennessee River valley. *Tennessee Anthropologist*, **6**, 61–84.

Chapman, J., and Watson, P. J. (1993). The Archaic Period and the flotation revolution. In *Foraging and Farming in the Eastern Woodlands*, ed. C. M. Scarry. Gainesville, FL: University Press of Florida, pp. 27–38.

Chapman, J., Delcourt, P. A., Cridlebaugh, P. A., Shea, A. B., and Delcourt, H. R. (1982). Man-land interaction: 10,000 years of American Indian impact on native ecosystems in the lower Little Tennessee River Valley, eastern Tennessee. *Southeastern Archaeology*, **1**, 115–21.

Clark, J. S. (1988a). Particle motion and the theory of charcoal analysis: source area, transport, deposition, and sampling. *Quaternary Research*, **30**, 81–91.

(1988b). Charcoal-stratigraphic analysis on petrographic thin sections: recent fire history in northwest Minnesota. *Quaternary Research*, **30**, 67–80.

Clark, J. S., and Royall, P. D. (1995). Transformation of a northern hardwood forest by aboriginal (Iroquois) fire: charcoal evidence from Crawford Lake, Ontario, Canada. *The Holocene*, **5**, 1–9.

Clements, F. (1936). Nature and structure of the climax. *Journal of Ecology*, **24**, 252–84.

Cohen, M. N. (1977). *The Food Crisis in Prehistory: Overpopulation and the Origins of Agriculture*. New Haven, CT: Yale University Press.

Colinvaux, P. A. (1978). *Why Big Fierce Animals are Rare: An Ecologist's Perspective.* Princeton, NJ: Princeton University Press.

Collins, M. B. (1998). Interpreting the Clovis artifacts from the Gault site. *TARL Research Notes, The Friends of the Texas Archeological Research Laboratory, The University of Texas at Austin*, **6**, 5–12.

Collins, M. B., Hester, T. R., Olmstead, D., and Headrick, P. J. (1991). Engraved cobbles from early archaeological contexts in Central Texas. *Current Research in the Pleistocene*, **8**, 13–15.

Connell, J. H. (1978). Diversity in tropical rain forests and coral reefs. *Science*, **199**, 1302–10.

Cook, N. D. (1998). *Born to Die: Disease and New World Conquest, 1492–1650.* Cambridge, UK: Cambridge University Press.

Cowan, C. W. (1985a). From foraging to incipient food production: subsistence change and continuity on the Cumberland Plateau of Eastern Kentucky. Ph.D. dissertation. Ann Arbor, MI: University Microfilms, University of Michigan.

(1985b). Understanding the evolution of plant husbandry in eastern North America: lessons from botany, ethnography and archaeology. In *Prehistoric Food Production in North America*, ed. R. I. Ford. Anthropological Papers No. 75, Museum of Anthropology. Ann Arbor, MI: University of Michigan, pp. 205–43.

(1997). Evolutionary changes associated with the domestication of *Cucurbita pepo*: evidence from eastern Kentucky. In *People, Plants, and Landscapes: Studies in Paleoethnobiology*, ed. K. J. Gremillion. Tuscaloosa, AL: University of Alabama Press, pp. 63–85.

Cowan, C. W., Jackson, H. E., Moore, K., Nickelhoff, A., and Smart, T. (1981). The Cloudsplitter Rockshelter, Menifee County, Kentucky: a preliminary report. *Southeastern Archaeological Conference Bulletin*, **24**, 60–75.

Crawford, G. W., Smith, D. G., and Bowyer, V. E. (1997). Dating the entry of corn (*Zea mays*) into the lower Great Lakes region. *American Antiquity*, **62**, 112–19.

Cridlebaugh, P. A. (1984). American Indian and Euro–American impact upon Holocene vegetation in the lower Little Tennessee River valley, east Tennessee. Ph.D. dissertation. Knoxville, TN: Department of Anthropology, University of Tennessee.

Crites, G. D. (1987). Human–plant mutualism and niche expression in the paleoethnobotanical record: a Middle Woodland example. *American Antiquity*, **52**, 725–40.

(1991). Investigations into early plant domestication and food production in Middle Tennessee: a status report. *Tennessee Anthropologist*, **16**, 69–87.

(1993). Domesticated sunflower in fifth millennium B.P. temporal context: new evidence from Middle Tennessee. *American Antiquity*, **58**, 146–8.

Cronon, W. (1983). *Changes in the Land: Indians, Colonists, and the Ecology of New England.* New York, NY: Hill and Wang.

Crosby, A. W. (1986). *Ecological Imperialism: The Biological Expansion of Europe, 900–1900.* Cambridge, UK: Cambridge University Press.

Davenport, C. D. (1999). Estimating sex and weight of *Odocoileus virginianus* (whitetail deer) with implications to human status at Toqua. M.S. thesis. Knoxville, TN: Department of Anthropology, University of Tennessee.

Davis, R. P. S., Jr. (1990). *Aboriginal Settlement Patterns in the Little Tennessee River Valley*, University of Tennessee Department of Anthropology Report of Investigations No. 50, Tennessee Valley Authority publications in Anthropology No. 54. Chattanooga, TN: Tennessee Valley Authority.

Day, G. M. (1953). The Indian as an ecological factor in the northeastern forest. *Ecology*, **34**, 329–46.

Decker, D. S., and Wilson, H. G. (1986). Numerical analysis of seed morphology in *Cucurbita pepo. Systematic Botany*, **11**, 595–607.

Decker-Walters, D. S. (1993). New methods for studying the origins of New World domesticates: the squash example. In *Foraging and Farming in the Eastern Woodlands*, ed. C. M. Scarry. Gainesville, FL: University Press of Florida, pp. 91–7.

Decker-Walters, D. S., Walters, T. W., Cowan, C. W., and Smith, B. D. (1993). Isozymic characterization of wild populations of *Cucurbita pepo. Journal of Ethnobiology*, **13**, 55–72.

Deevey, E. S. (1969). Coaxing history to conduct experiments. *BioScience*, **19**, 40–3.

Delcourt, H. R. (1987). The impact of prehistoric agriculture and land occupation on natural vegetation. *Trends in Ecology and Evolution*, **2**, 39–44.

(2002). *Forests in Peril: Tracking Deciduous Trees from Ice-Age Refuges into the Greenhouse World.* Blacksburg, VA: McDonald & Woodward Publishing Company.

Delcourt, H. R., and Delcourt, P. A. (1988). Quaternary landscape ecology: relevant scales in space and time. *Landscape Ecology*, **2**, 23–44.

(1991). *Quaternary Ecology: A Paleoecological Perspective.* New York, NY: Chapman and Hall.

(1994). Postglacial rise and decline of *Ostrya virginiana* (Mill.) K. Koch and *Carpinus caroliniana* Walt. in eastern North America: predictable responses of forest species to cyclic changes in seasonality of climates. *Journal of Biogeography*, **21**, 137–50.

(1997). Pre-Columbian Native American use of fire on southern Appalachian landscapes. *Conservation Biology*, **11**, 1010–14.

(2000). Eastern deciduous forests. In *North American Terrestrial Vegetation,* 2nd edition, ed. M. G. Barbour and D. W. Billings. Cambridge, UK: Cambridge University Press, pp. 357–95.

Delcourt, H. R., Delcourt, P. A., and Royall, P. D. (1997). Late Quaternary vegetational history of the Western Lowlands. In *Sloan: A Paleoindian Dalton Cemetery in Arkansas*, ed. D. F. Morse. Washington, D.C.: Smithsonian Institution Press, pp. 103–22.

Delcourt, H. R., Delcourt, P. A., and Webb, T., III. (1983). Dynamic plant ecology: the spectrum of vegetational change in space and time. *Quaternary Science Reviews*, **1**, 153–75.

Delcourt, P. A. (1980a). Goshen Springs: late Quaternary vegetation record for southern Alabama. *Ecology*, **61**, 371–86.

(1980b). Quaternary alluvial terraces of the Little Tennessee River Valley, East Tennessee. In *The 1979 Archaeological and Geological Investigations in the Tellico Reservoir*, ed. J. Chapman, Report of Investigations, No. 29, pp. 110–21, 175–212.

Delcourt, P. A., and H. R. Delcourt. (1984). Late-Quaternary paleoclimates and biotic responses in eastern North America and the western North Atlantic Ocean. *Palaeogeography, Palaeoclimatology, Palaeoecology*, **48**, 263–84.

(1987). *Long-term Forest Dynamics of Eastern North America, Ecological Studies 63.* New York, NY: Springer-Verlag.

(1993). Paleoclimates, paleovegetation, and paleofloras during the late Quaternary. In *Flora of North America North of Mexico. Volume 1: Introduction*, ed. N. R. Morin, New York, NY: Oxford University Press, pp. 71–94.

(1996). Quaternary paleoecology of the Lower Mississippi Valley. *Engineering Geology*, **45**, 219–42.

(1998a). The influence of prehistoric human-set fires on oak-chestnut forests in the southern Appalachians. *Castanea*, **63**, 337–45.

(1998b). Paleoecological insights on conservation of biodiversity: a focus on species, ecosystems, and landscapes. *Ecological Applications*, **8**, 921–34.

Delcourt, P. A., Delcourt, H. R., Cridlebaugh, P. A., and Chapman, J. (1986). Holocene ethnobotanical and paleoecological record of human impact on vegetation in the Little Tennessee River Valley, Tennessee. *Quaternary Research*, **25**, 330–49.

Delcourt, P. A., Delcourt, H. R., Morse, D. F., and Morse, P. A. (1993). History, evolution, and organization of vegetation and human culture. In *Biodiversity of the Southeastern United States*, ed. W. H. Martin, S. G. Boyce, and A. C. Echternacht, Vol. 1. New York, NY: John Wiley & Sons, pp. 47–79.

Delcourt, P. A., Delcourt, H. R., Ison, C. R., Sharp, W. E., and Gremillion, K. J. (1998). Prehistoric human use of fire, the Eastern Agricultural Complex, and Appalachian oak–chestnut forests: paleoecology of Cliff Palace Pond, Kentucky. *American Antiquity*, **63**, 263–78.

Delcourt, P. A., Delcourt, H. R., and Saucier, R. T. (1999). Late Quaternary vegetation dynamics in the Central Mississippi Valley. In *Arkansas Archeology: Essays in Honor of Dan and Phyllis Morse*, ed. R. C. Mainfort, Jr., and M. D. Jeter. Fayetteville, AR: University of Arkansas Press, pp. 15–30.

Delcourt, P. A., Nester, P. L., Delcourt, H. R., Mora, C. I., and Orvis, K. H. (2002). Holocene lake-effect precipitation in Northern Michigan. *Quaternary Research*, **57**, 225–33.

Denevan, W. M. (1992). The pristine myth: the landscape of the Americas in 1492. *Annals of the Association of American Geographers*, **82**, 369–85.

Denslow, J. S. (1980). Patterns of plant species diversity during succession under different disturbance regimes. *Oecologia*, **46**, 18–21.

(1985). Disturbance-mediated coexistence of species. In *The Ecology of Natural Disturbance and Patch Dynamics*, ed. S. T. A. Pickett and P. S. White. Orlando, FL: Academic Press, pp. 307–23.

Densmore, F. 1927. Uses of plants by the Chippewa Indians. *Bureau of American Ethnology 44th Annual Report*, 275–397.

Desowitz, R. S. (1997). *Who gave Pinta to the Santa Maria? Torrid Diseases in a Temperate World.* San Diego, CA: Harvest Books, Harcourt Brace & Company.

Dey, D. C., and Guyette, R. P. (2000). Anthropogenic fire history and red oak forests in south-central Ontario. *The Forestry Chronicle*, **76**, 339–47.

Diamond, J. M. (1984). Historic extinctions: a Rosetta Stone for understanding prehistoric extinctions. In *Quaternary Extinctions: A Prehistoric Revolution*, ed. P. S. Martin and R. G. Klein. Tucson, AZ: University of Arizona Press, pp. 824–62.

(1986). The environmentalist myth: archaeology. *Nature*, **324**, 19–20.

Dillehay, T. D. (1991). Disease ecology and initial human migration. In *The First Americans: Search and Research*, ed. T. D. Dillehay, and D. J. Meltzer. Boca Raton, FL: CRC Press, pp. 209–30.

(1997). *Monte Verde: A Late Pleistocene Settlement in Chile. Volume 2: The Archaeological Context.* Washington, DC: Smithsonian Institution Press.

(2000). *The Settlement of the Americas: A New Prehistory.* New York, NY: Basic Books.

Dixon, E. J. (1999). *Bones, Boats, & Bison, Archaeology and the First Colonization of Western North America.* Albuquerque, NM: The University of New Mexico Press.

Dobyns, H. F. (1983). *Their Number Become Thinned: Native American Populations Dynamics in Eastern North America.* Knoxville, TN: University of Tennessee Press.

Doebley, J. F. (1994). Morphology, molecules, and maize. In *Corn and Culture in the Prehistoric New World*, ed. S. Johannessen and C. A. Hastorf. Boulder, CO: Westview Press, pp. 101–12.

Dolph, G. E. (1993). The myth of the natural man. *Indiana Academy of Science*, **102**, 237–46.

Doran, G. H., Dickel, D. N., and Newsom, L. A. (1990). A 7,290-year-old bottle gourd from the Windover Site, Florida. *American Antiquity*, **55**, 354–9.

Driver, H. E., and Massey, W. C. (1957). Comparative studies of North American Indians. *Transactions of the American Philosophical Society, n.s.*, **47**, 165–449.

Drooker, P. B. (1992). *Mississippian Village Textiles at Wickliffe.* Tuscaloosa, AL: University of Alabama Press.

Dunning, J. B., Danielson, B. J., and Pulliam, H. R. (1992). Ecological processes that affect populations in complex landscapes. *Oikos*, **65**, 169–75.

Dyke, A. S., and Prest, V. K. (1987). Late Wisconsinan and Holocene history of the Laurentide Ice Sheet. *Géographie physique et Quaternaire*, **XLI**, 237–63.

Egler, F. E. (1977). *The Nature of Vegetation, Its Management and Mismanagement.* Norfolk, CT: Aton Forest and Connecticut Conservation Association.

Elias, S. A. (2000). Late Pleistocene climates of Beringia, based on analysis of fossil beetles. *Quaternary Research*, **53**, 229–35.

Ellis, C., Goodyear, A. C., Morse, D. F., and Tankersley, K. B. (1998). Archaeology of the Pleistocene–Holocene transition in Eastern North America. *Quaternary International*, **49/50**, 151–66.

Esarey, D., and Pauketat, T. R. (1992). The Lohmann Site: an early Mississippian center in the American Bottom (11-S-49). *American Bottom Archaeology*, FAI-270 Site Reports, Vol. 25. Urbana, IL: University of Illinois Press.

Fagan, W. F., Meir, E., and Moore, J. L. (1999). Variation thresholds for extinction and their implications for conservation strategies. *The American Naturalist*, **154**, 510–20.

Ferguson, S. H., Taylor, M. K., and Messier, F. (2000). Influence of sea ice dynamics on habitat selection by polar bears. *Ecology*, **81**, 761–72.

Fladmark, K. R. (1979). Routes: alternative migration corridors for early man in North America. *American Antiquity*, **44**, 55–69.

(1990). Possible early human occupation in the Queen Charlotte Island. *Canadian Journal of Archaeology*, **14**, 183–97.

Ford, J., and Martinez, D. (2000). Invited feature: traditional ecological knowledge, ecosystem science, and environmental management. *Ecological Applications*, **10**, 1249–50.

Ford, R. I. (1979). Paleoethnobotany in American archaeology. *Advances in Archaeological Method and Theory*, **2**, 285–336.

(1981). Gardening and farming before A.D. 1000: patterns of prehistoric cultivation north of Mexico. *Journal of Ethnobiology*, **1**, 6–27.

(1985). The processes of plant food production in prehistoric North America. In *Prehistoric Food Production in North America.* Anthropological Papers No. 75, ed. R. I. Ford, Ann Arbor, MI: Museum of Anthropology, University of Michigan, pp. 1–18.

Forman, R. T. T., and Russell, E. W. B. (1983). Evaluation of historical data in ecology. *Bulletin of the Ecological Society of America*, **64**, 5–7.

Forster, P., Harding, R., Torroni, A., and Bandelt, H.-J. (1996). Origin and evolution of Native American mtDNA variation: a reappraisal. *American Journal of Human Genetics*, **59**, 935–45.

Fowler, M. L. (1989). *The Cahokia Atlas: A Historical Atlas of Cahokia Archaeology, Studies in Illinois Archaeology No. 6.* Springfield, IL: Illinois Historic Preservation Agency.

Fowler, M L., and Hall, R. L. (1975). Archaeological phases at Cahokia. *Perspectives in Cahokia Archaeology*, Illinois Archaeological Survey Bulletin No. 10, 1–14.

Franklin, J. F. (1993). Preserving biodiversity: species, ecosystems, or landscapes? *Ecological Applications*, **3**, 202–5.

Fritz, G. J. (1990). Multiple pathways to farming in precontact eastern North America. *Journal of World Prehistory*, **4**, 387–435.

(1995). New dates and data on early agriculture: the legacy of complex hunter–gatherers. *Annals of the Missouri Botanical Garden*, **82**, 3–15.

(1999). Gender and the early cultivation of gourds in eastern North America. *American Antiquity*, **64**, 417–29.

(2000). Levels of native biodiversity in eastern North America. In *Biodiversity and Native America*, ed. P. E. Minnis and W. J. Elisens. Norman, OK: University of Oklahoma Press, pp. 223–47.

Gardner, P. S. (1997). The ecological structure and behavioral implications of mast exploitation strategies. In *People, Plants, and Landscapes: Studies in Paleoethnobotany*, ed. K. J. Gremillion. Tuscaloosa, AL: University of Alabama Press, pp. 161–78.

Gardner, R. H., and R. V. O'Neill. (1991). Pattern, process, and predictability: the use of neutral models for landscape analysis. In *Quantitative Methods in Landscape Ecology, Ecological Studies 82*, ed. M. G. Turner and R. H. Gardner. New York, NY: Springer-Verlag, pp. 289–307.

Gardner, R. H., Milner, B. T., Turner, M. G., and O'Neill, R. V. (1987). Neutral models for the analysis of broad-scale landscape patterns. *Landscape Ecology*, **1**, 19–28.

Gauch, H. G., Jr. (1982). *Multivariate Analysis in Community Ecology*. Cambridge, UK: Cambridge University Press.

Geist, V. (1999). Periglacial ecology, large mammals, and their significance to human biology. In *Ice Age People of North America: Environments, Origins, and Adaptations*, ed. R. Bonnischen, and K. L. Turnmire. Corvallis, OR: Oregon State University Press and the Center for the Study of the First Americans, pp. 78–94.

Gillam, J. C. (1999). Paleoindian settlement in northeastern Arkansas. In *Arkansas Archaeology: Essays in Honor of Dan and Phyllis Morse*, ed. R. C. Mainfort, Jr., and M. D. Jeter. Fayetteville, AR: University of Arkansas Press, pp. 99–118.

Golley, F. B. (1984). Historical origins of the ecosystem concept in ecology. In *Ecosystem Concept in Ecology*, ed. E. Moran. Washington, D.C.: American Association for the Advancement of Science Publications, pp. 33–49.

Goloubinoff, P., Pääbo, S., and Wilson, A. C. (1994). Molecular characterization of ancient maize: potentials and pitfalls. In *Corn and Culture in the Prehistoric New World*, ed. S. Johannessen and C. A. Hastorf. Boulder, CO: Westview Press, pp. 113–25.

González, J. J. S. (1994). Modern variability and patterns of maize movement in Mesoamerica. In *Corn and Culture in the Prehistoric New World*, ed. S. Johannessen and C. A. Hastorf. Boulder, CO: Westview Press, pp. 135–56.

Goodyear, A. C. (1982). The chronological position of the Dalton Horizon in the southeastern United States. *American Antiquity*, **47**, 82–95.

Graham, R. W. (1986). Response of mammalian communities to environmental changes during the Late Quaternary. In *Community Ecology*, ed. J. Diamond and T. J. Case. New York, NY: Harper and Row, pp. 300–13.

(1990). Evolution of new ecosystems at the end of the Pleistocene. In *Megafauna and Man: Discovery of America's Heartland, Scientific Papers of the Mammoth Site of Hot Springs, South Dakota, Inc.*, ed. L. D. Agenbroad, J. I. Mead, and L. W. Nelson, Vol. 1. Hot Springs, SD: Mammoth Site of Hot Springs, SD, Inc, pp. 53–60.

Graham, R. W., and Lundelius, E. L., Jr. (1984). Coevolutionary disequilibrium and Pleistocene extinctions. In *Quaternary Extinctions: A Prehistoric Revolution,* ed. P. S. Martin and R. G. Klein. Tucson, AZ: University of Arizona Press, pp. 223–49.

(1994). FAUNMAP: a database documenting late Quaternary distributions of mammal species in the United States. *Illinois State Museum Scientific Papers,* vol. XXV, No. 1 & 2. Springfield, IL: Illinois State Museum (web site at http://www.museum.state.il.us/research/faunmap/).

Graham, R. W., and Mead, J. I. (1987). Environmental fluctuations and evolution of mammalian faunas during the last deglaciation in North America. In *North America and Adjacent Oceans during the Last Deglaciation, the Geology of North America, The Decade of North American Geology Vol. K-3,* ed. W. F. Ruddiman and H. E. Wright, Jr. Boulder, CO: Geological Society of America, pp. 371–402.

Grayson, D. K. (1989). The chronology of North American late Pleistocene extinctions. *Journal of Archaeological Science,* **16**, 153–65.

(1991). Late Pleistocene mammalian extinctions in North America: taxonomy, chronology, and explanations. *Journal of World Prehistory,* **5**, 193–231.

(2003). Reassessing overkill: early Americans and Pleistocene mammals. In *Zooarchaeology: Papers in Honor of Elizabeth S. Wing,* ed. C. M. Porter, vol. 43. Gainesville, FL: *Bulletin of the Florida Museum of Natural History.*

Greenlee, D. M. (1998). Prehistoric diet in the central Mississippi River valley. In *Changing Perspectives on the Archaeology of the Central Mississippi River Valley,* ed. M. J. O'Brien and R. C. Dunnell, Tuscaloosa, AL: University of Alabama Press, pp. 299–324.

Gregg, M. L. (1975). A population estimate for Cahokia. *Perspectives in Cahokia Archaeology, Illinois Archaeological Survey Bulletin,* **10**, 126–36.

Gremillion, K. J. (1993a). Crop and weed in prehistoric eastern North America: the *Chenopodium* example. *American Antiquity,* **58**, 496–508.

(1993b). Plant husbandry at the Archaic/Woodland transition: evidence from the Cold Oak Shelter, Kentucky. *Midcontinental Journal of Archaeology,* **18**, 161–89.

(1996). Early agricultural diet in eastern North America: evidence from two Kentucky rockshelters. *American Antiquity,* **61**, 520–36.

(1997). New perspectives on the paleoethnobotany of the Newt Kash Shelter. In *People, Plants, and Landscapes: Studies in Paleoethnobotany,* ed. K. J. Gremillion. Tuscaloosa, AL: University of Alabama Press, pp. 23–41.

(1999). *National Register Evaluation of the Courthouse Rock Shelter (15PO322), Powell County, Kentucky.* Winchester, KY: USDA Forest Service, Daniel Boone National Forest.

Gremillion, K. J., and Ison, C. R. (1992). Terminal Archaic and Early Woodland plant utilization at the Cold Oak Shelter. In *Upland Archaeology in the East: Symposium IV,* ed. M. B. Barber and E. B. Barfield. Cultural Resource Management Report No. 92–1. Atlanta, GA: USDA Forest Service Southern Region, pp. 121–32.

Gremillion, K. J., and Sobolik, K. D. (1996). Dietary variability among prehistoric forager–farmers of eastern North America. *Current Anthropology*, **37**, 529–39.

Griffin, J. B. (1952). *Archaeology of the Eastern United States*. Chicago, IL: University of Chicago Press.

(1978). Foreword. In *Mississippian Settlement Patterns*, ed. B. D. Smith. New York, NY: Academic Press, pp. xv–xxii.

(1984). A historical perspective. In *American Bottom Archaeology: A Summary of the FAI-270 Project Contribution to the Culture History of the Mississippi River Valley*, ed. C. J. Bareis and J. W. Porter. Urbana, IL: University of Illinois Press, pp. xv–xviii.

(1985). Changing concepts of the prehistoric Mississippian cultures of the eastern United States. In *Alabama and the Borderlands: From Prehistory to Statehood*, ed. R. Badger and L. Clayton. Tuscaloosa, AL: University of Alabama Press, pp. 40–63.

Grime, J. P. (1979). *Plant Strategies and Vegetation Processes*. New York, NY: John Wiley & Sons.

Grimm, E. C. (1983). Chronology and dynamics of vegetation change in the prairie–woodland region of southern Minnesota, USA. *New Phytologist*, **93**, 311–50.

Gunderson, L., Holling, C. S., and Light, S. (eds.). (1995). *Barriers and Bridges to the Renewal of Ecosystems and Institutions*. New York, NY: Columbia University Press.

Gunderson, L. H., and Holling, C. S. (eds.). (2001). *Panarchy: Understanding Transformations in Human and Natural Systems*. Washington, DC: Island Press.

Guthrie, R. D. (1984). Mosaics, allochemics, and nutrients: an ecological theory of Late Pleistocene megafaunal extinctions. In *Quaternary Extinctions: A Prehistoric Revolution*, ed. P. S. Martin, and R. G. Klein, Tucson, AZ: The University of Arizona Press, pp. 259–98.

Hamel, P. B., and Buckner, E. R. (1998). How far could a squirrel travel in the treetops? A prehistory of the southern forest. *Transactions of the 63rd North American Wildlife and Natural Resources Conference*, pp. 309–15.

Hammett, J. E. (1992). Ethnohistory of aboriginal landscapes in the southeastern United States. *Southern Indian Studies*, **41**, 1–50.

(1997). Interregional patterns of land use and plant management in native North America. In *People, Plants, and Landscapes: Studies in Paleoethnobotany*, ed. K. J. Gremillion, Tuscaloosa, AL: University of Alabama Press, pp. 195–216.

(2000). Ethnohistory of aboriginal landscapes in the southeastern United States. In *Biodiversity and Native America*, ed. P. E. Minnis and W. J. Elisens. Norman, OK: University of Oklahoma Press, pp. 248–99.

Harris, D. R. (1989). An evolutionary continuum of people–plant interaction. In *Foraging and Farming: The Evolution of Plant Exploitation*, ed. D. R. Harris and G. C. Hillman. London, UK: Unwin Hyman Ltd, pp. 11–26.

Hart, J. P., and Scarry, C. M. (1999). The age of common beans (*Phaseolus vulgaris*) in the northeastern United States. *American Antiquity*, **64**, 653–8.

Hayden, B. (1992). Models of domestication. In *Transitions to Agriculture in Prehistory*, ed. A. B. Gebauer and T. D. Price. Madison, WI: Prehistory Press, pp. 11–19.

Haynes, C. V., Jr. (1991). Geoarchaeological and paleohydrological evidence for a Clovis-age drought in North America and its bearing on extinction. *Quaternary Research*, **35**, 438–50.

(1995). Geochronology of paleoenvironmental change, Clovis Type Site, Blackwater Draw, New Mexico. *Geoarchaeology: An International Journal*, **10**, 317–88.

(2000). New World climate: dramatic climatic shifts welcomed and bedeviled the First Americans. *Scientific American Discovering Archaeology*, **2**, 37–9.

Haynes, G. (1991). *Mammoths, Mastodonts, and Elephants*. Cambridge, UK: Cambridge University Press.

Heiser, C. B., Jr. (1989). Domestication of Cucurbitaceae: *Cucurbita* and *Lagenaria*. In *Foraging and Farming: The Evolution of Plant Exploitation*, ed. D. R. Harris and G. C. Hillman. London, UK: Unwin Hyman Ltd, pp. 471–80.

Henige, D. (1998). *Numbers from Nowhere: The American Indian Contact Population Debate*. Norman, OK: University of Oklahoma Press.

Henderson, D., and Hedrick, L. D. (1991). *Restoration of Old Growth Forests in the Interior Highlands of Arkansas and Oklahoma*. Proceedings of Winrock International Conference, September 19–20, 1990. Morrilton, AR: Winrock International Institute for Agricultural Development.

Heusser, C. J. (1989). North Pacific coastal refugia – The Queen Charlotte Islands in perspective. In *The Outer Shores*, ed. G. G. E. Scudder, and N. Gessler. Skidegate, BC: Queen Charlotte Islands Museum, pp. 91–106.

Holliday, V. T. (2000). Folsom drought and episodic drying on the Southern High Plains from 10,900–10,200 ^{14}C yr B.P. *Quaternary Research*, **53**, 1–12.

Holling C. S. (1978). *Adaptive Environmental Assessment and Management*. London: John Wiley.

(1995). What barriers? What bridges? In *Barriers and Bridges to the Renewal of Ecosystems and Institutions*, ed. L. H. Gunderson, C. S. Holling, and S. S. Light. New York, NY: Columbia University Press, pp. 3–34.

(2001). Understanding the complexity of economic, ecological, and social systems. *Ecosystems*, **4**, 390–405.

Hudson, C. (1997). *Knights of Spain, Warriors of the Sun: Hernando de Soto and the South's Ancient Chiefdoms*. Athens, GA: University of Georgia Press.

Hughes, T., Borns, H. W., Jr., Fastook, *et al.* (1985). Models of glacial reconstruction and deglaciation applied to maritime Canada and New England. In *Geological Society of America Special Paper 197*, ed. H. W. Borns, Jr., P. LaSalle, and W. B. Thompson, pp. 139–50.

Hunter, M. L., Jr., Jacobson, G. L., Jr., and Webb, T., III. (1988). Paleoecology and the coarse-filter approach to maintaining biological diversity. *Conservation Biology*, **2**, 375–85.

Huntley, B., and Birks, H. J. B. (1983). *An Atlas of Past and Present Pollen Maps for Europe: 0–13 000 Years Ago*. Cambridge, UK: Cambridge University Press.

Hurd, P. D., Jr., Linsley, E. G., and Whitaker, T. W. (1971). Squash and gourd bees (*Peponapis, Xenoglossa*) and the origin of the cultivated *Cucurbita*. *Evolution*, **25**, 218–34.

Imbrie, J., and Imbrie, K. (1979). *Ice Ages, Solving the Mystery*. Hillside, NJ: Enslow.

Imbrie, J., McIntyre, A., and Moore, T. C., Jr. (1983). The ocean around North America at the last glacial maximum. In *Late-Quaternary Environments of the United States,* ed. S. C. Porter, Vol. 1. Minneapolis, MN: University of Minnesota Press, pp. 230–6.

Ison, C. R. (1988). The Cold Oak Shelter: providing a better understanding of the terminal archaic. In *Paleoindian and Archaic Research in Kentucky*, ed. C. D. Hockensmith, D. Pollack, and T. N. Sanders, Frankfort, KY: Kentucky Heritage Council, pp. 205–20.

(1991). Prehistoric upland farming along the Cumberland Plateau. In *Studies in Kentucky Archaeology*, ed. C. D. Hockensmith, Frankfort, KY: Kentucky Heritage Council, pp. 1–10.

(1996). *Daniel Boone National Forest Radiocarbon Dates*. Winchester, KY: USDA Forest Service, Daniel Boone National Forest.

Iversen, J. (1958). The bearing of glacial and interglacial epochs on the formation and extinction of plant taxa. *Uppsala Universiteit Arssk*, **6**, 210–15.

Jackson, S. T., and Weng, C. (1999). Late Quaternary extinction of a tree species in eastern North America. *Proceedings of the National Academy of Sciences of the United States of America*, **96**, 13847–52.

Jacobson, G. L., Jr., Webb, T., III, and Grimm, E. C. (1987). Patterns and rates of vegetation change during the deglaciation of eastern North America. In *North America and Adjacent Oceans during the Last Deglaciation, the Geology of North America, The Decade of North American Geology Vol. K-3*, ed. W. F. Ruddiman and H. E. Wright, Jr. Boulder, CO: Geological Society of America, pp. 277–88.

Jakes, K. A., and Sibley, L. R. (1994). A comparative collection for the study of fibers used in prehistoric textiles from eastern North America. *Journal of Archaeological Science*, **21**, 641–50.

Janzen, D. H. (1969). Seed-eaters versus seed size, number, toxicity, and dispersal. *Evolution*, **23**, 1–27.

Janzen, D. H., and Martin, P. S. (1982). Neotropical anachronisms: the fruits the Gomphotheres ate. *Science*, **215**, 19–27.

Johannessen, S. (1984). Paleoethnobotany. In *American Bottom Archaeology*, ed. C. J. Bareis and J. W. Porter. Urbana, IL: University of Illinois Press, pp. 197–214.

Johnson, A. W., and Earle, T. (2000). *The Evolution of Human Societies: From Foraging Group to Agrarian State,* 2nd edn. Stanford, CA: Stanford University Press.

Kaplan, L. (1981). What is the origin of the common bean? *Economic Botany*, **35**, 240–54.

Kapp. R. O. (1977). Late Pleistocene and postglacial plant communities of the Great Lakes region. In *Geobotany*, ed. R. Romans. New York, NY: Plenum, pp. 1–27.

Kay, C. E. (1995). Aboriginal overkill and native burning; implications for modern ecosystem management. *Western Journal of Applied Forestry*, **10**, 121–6.

Kay, M., King, F. B., and Robinson, C. K. (1980). Cucurbits from Phillips Spring: new evidence and interpretations. *American Antiquity*, **45**, 806–22.

Kelly, J. E. (1991). Cahokia and its role as a gateway center in interregional exchange. In *Cahokia and the Hinterlands: Middle Mississippian Cultures of the Midwest*, ed. T. E. Emerson and R. B. Lewis. Urbana, IL: University of Illinois Press, pp. 61–80.

Kelly, J. E., Ozuk, S. J., Jackson, D. K., *et al.* (1984). Emergent Mississippian period. In *American Bottom Archaeology*, ed. C. J. Bareis and J. W. Porter. Urbana, IL: University of Illinois Press, pp. 128–57.

Kelly, L. S., and Cross, P. G. (1984). Zooarchaeology. In *American Bottom Archaeology*, ed. C. J. Bareis and J. W. Porter. Urbana, IL: University of Illinois Press, pp. 215–32.

Kelly, R. (1995). *The Forager Spectrum*. Washington, DC: Smithsonian Press.

Kelly, R. L., and Todd, L. C. (1988). Coming into the country: Early Paleoindian hunting and mobility. *American Antiquity*, **53**, 231–44.

Kimball, L. R. (1985). *The 1977 Archaeological Survey: An Overall Assessment of the Archaeological Resources of Tellico Reservoir,* Report of Investigations No. 40. Knoxville, TN: Department of Anthropology, University of Tennessee.

Kimmerer, R. W. (2000). Native knowledge for native ecosystems. *Journal of Forestry*, **98**, 4–9.

King, J. E., and Saunders, J. J. (1984). Environmental insularity and the extinction of the American mastodont. In *Quaternary Extinctions: A Prehistoric Revolution*, ed. P. S. Martin and R. G. Klein. Tucson, AZ: University of Arizona Press, pp. 315–39.

Kloor, K. (2000). Returning America's forests to their natural roots. *Science*, **287**, 573–5.

Knox, J. C. (1985). Responses of floods to Holocene climatic change in the upper Mississippi valley. *Quaternary Research*, **23**, 287–300.

(1996). Late Quaternary upper Mississippi River alluvial episodes and their significance to the lower Mississippi River system. *Engineering Geology*, **45**, 263–85.

Kohler, T. A. (1992). Prehistoric human impact on the environment in the upland North American Southwest. *Population and Environment: A Journal of Interdisciplinary Studies*, **13**, 255–68.

Kohler, T. A., Kresl, J., Van West, C., Carr, E., and Wilshusen, R. H. (2000). Be there then: a modeling approach to settlement determinants and spatial efficiency among late ancestral Pueblo populations of the Mesa Verde Region, U.S. Southwest. In *Dynamics in Human and Primate Societies: Agent-Based Modeling of Social and Spatial Processes*, ed. T. A. Kohler and G. Gumerman. New York, NY: Oxford University Press, pp. 145–78.

Kot, M., Lewis, M. A., and van den Driessche, P. (1996). Dispersal data and the spread of invading organisms. *Ecology*, **77**, 2027–42.

Kowalewski, S. A. (1995). Large-scale ecology in aboriginal eastern North America. In *Native American Interactions: Multiscalar Analyses and Interpretations in the*

Eastern Woodlands, ed. M. S. Nassaney and K. E. Sassaman. Knoxville, TN: University of Tennessee Press, pp. 147–73.
Krech, S., III. (1999). *The Ecological Indian, Myth and History*. New York, NY: W. W. Norton & Company.
Küchler, A. W. (1964). *The Potential Natural Vegetation of the Conterminous United States*. Special Publication, no. 36. New York, NY: American Geographical Society.
Kurtén, B., and Anderson, E. (1980). *Pleistocene Mammals of North America*. New York, NY: Columbia University Press.
Kuttruff, J. T., DeHart, S. G., and O'Brien, M. J. (1998). 7500 years of prehistoric footwear from Arnold Research Cave, Missouri. *Science*, **281**, 72–5.
Kutzbach, J. E., and Guetter, P. J. (1986). The influence of changing orbital parameters and surface boundary conditions on climate simulations for the past 18,000 years. *Journal of the Atmospheric Sciences*, **43**, 1726–59.
Lambert, W. D., and Holling, C. S. (1998). Causes of ecosystem transformation at the end of the Pleistocene: evidence from mammal body-mass distributions. *Ecosystems*, **1**, 157–175.
Lengyel, S. N., Eighmy, J. L., and Sullivan, L. P. (1999). On the potential of archaeomagnetic dating in the midcontinent region of North America: Toqua site results. *Southeastern Archaeology*, **18**, 156–71.
Levin, S. A. (1992). The problem of pattern and scale in ecology. *Ecology*, **73**, 1943–67.
Lockwood, J. G. (1979) *Causes of Climate*. New York, NY: John Wiley and Sons.
Lodge, D. M. (1993a). Biological invasions: lessons for ecology. *Trends in Ecology and Evolution*, **8**, 133–7.
(1993b). Species invasions and deletions: community effects and responses to climate and habitat change. In *Biotic Interactions and Global Change*, ed. P. M. Karieva, J. G. Kingsolver, and R. B. Huey. Sunderland, MA: Sinauer Associates, pp. 367–87.
Lopinot, N. H., and Woods, W. I. (1993). Wood overexploitation and the collapse of Cahokia. In *Foraging and Farming in the Eastern Woodlands*, ed. C. M. Scarry. Gainesville, FL: University Press of Florida, pp. 206–31.
Low, T. (1999). *Feral Future: The Untold Story of Australia's Exotic Invaders*. Victoria, Australia: Viking Press.
Lowe, I. J., Ammann, B., Birks, H. H. *et al.* (1994). Climatic changes in areas adjacent to the North Atlantic during the last glacial–interglacial transition (14-9 Ka B.P): a contribution to IGCP-253. *Journal of Quaternary Science*, **9**, 185–98.
Loy, T. H., and Dixon, E. J. (1998). Blood residues on fluted points from eastern Beringia. *American Antiquity*, **63**, 21–46.
Lubchenco, J. (1978). Plant species diversity in a marine intertidal community: importance of herbivore food preference and algal competitive abilities. *American Naturalist*, **112**, 23–39.
Lubchenco, J., Olson, A. M., Brubaker, L. B., *et al.* (1991). The sustainable biosphere initiative: an ecological research agenda. *Ecology*, **72**, 371–412.

Lundelius, E. L., Jr., Graham, R. W., Anderson, E., *et al.* (1983). Terrestrial vertebrate faunas. In *Late-Quaternary Environments of the United States. Volume 1: The Late Pleistocene*, ed. S. C. Porter. Minneapolis, MN: University of Minnesota Press, pp. 311–53.

Lynott, M. J., Boutton, T. W., Price, J. E., and Nelson, D. E. (1986). Stable carbon isotopic evidence for maize agriculture in southeast Missouri and northeast Arkansas. *American Antiquity*, **51**, 51–65.

McAndrews, J. H. (1988). Human disturbance of North American forests and grasslands: the fossil pollen record. In *Vegetation History*, ed. B. Huntley and T. Webb, III. Dordrecht, Netherlands: Kluwer, pp. 673–97.

McAndrews, J. H., and Boyko-Diakonow, M. (1989). Pollen analysis of varved sediment at Crawford Lake, Ontario: evidence of Indian and European farming. In *Quaternary Geology of Canada and Greenland*, ed. R. J. Fulton. Ottawa, Ontario: Geological Survey of Canada, pp. 528–30.

MacArthur, R. H., and Pianka, E. (1966). On optimal use of a patchy environment. *The American Naturalist*, **100**, 603–9.

MacPhee, R. D. E., and Marx, P. A. (1997). The 40,000-year plague: humans, hyperdisease, and first-contact extinctions. In *Natural Change and Human Impact in Madagascar*, ed. S. M. Goodman and B. D. Patterson. Washington, D.C.: Smithsonian Institution Press, pp. 169–217.

McIntosh, R. P. (1985). *The Background of Ecology: Concept and Theory.* Cambridge, UK: Cambridge Studies in Ecology, Cambridge University Press.

McLauchlan, K. (2003). Plant cultivation and forest clearance by prehistoric North Americans: pollen evidence from Fort Ancient, Ohio, USA. *The Holocene*, **13**, 557–66.

McMillan, R. B. (1976). The dynamics of cultural and environmental change at Rodgers Shelter, Missouri. In *Prehistoric Man and His Environments: A Case Study in the Ozark Highland*, ed. W. R. Wood and R. B. McMillan. New York, NY: Academic Press, pp. 211–32.

Mack, R. N. (1985). Invading plants: their potential contribution to population biology. In *Studies in Plant Demography*, ed. J. White. London, UK: Academic Press, pp. 127–42.

McNaughton, S. J., and Georgiadis, N. J. (1986). Ecology of African grazing and browsing mammals. *Annual Review of Ecology and Systematics*, **17**, 39–65.

Mandryk, C. A. S., Josenhans, H., Fedje, D. W., and Mathewes, R. W. (2001). Late Quaternary paleoenvironments of northwestern North America: implications for inland versus coastal migration routes. *Quaternary Science Reviews*, **20**, 301–14.

Marks, P. L. (1983). On the origin of the field plants of the northeastern United States. *American Naturalist*, **122**, 210–28.

Marschner, F. J. (1959). *Land Use and Its Patterns in the United States*, Agricultural Handbook, No. 153. Washington, D.C.: United States Department of Agriculture.

(1974). *The Original Vegetation of Minnesota*, (map). St. Paul, MN: United States Forest Service, North Central Forest Experiment Station.

Martin, P. S. (1958). Pleistocene ecology and biogeography of North America. In *Zoogeography*, ed. C. L. Hubbs. Washington, D.C.: American Association for the Advancement of Science, pp. 375–420.

(1963). *The Last 10,000 Years*. Tucson, AZ: University of Arizona Press.

(1967). Prehistoric overkill. In *Pleistocene Extinctions: The Search for a Cause*, ed. P. S. Martin and H. E. Wright, Jr. New Haven, CT: Yale University Press, pp. 75–120.

(1984). Prehistoric overkill: the global model. In *Quaternary Extinctions: A Prehistoric Revolution*, ed. P. S. Martin and R. G. Klein. Tucson, AZ: The University of Arizona Press, pp. 354–403.

Mayewski, P. A., Denton, G. H., and Hughes, T. J. (1981). Late Wisconsin ice sheets in North America. In *The Last Great Ice Sheets*, ed. G. H. Denton and T. J. Hughes. New York, NY: John Wiley & Sons, pp. 67–178.

Mehrer, M. W., and Collins, J. M. (1995). Household archaeology at Cahokia and in its hinterlands. In *Mississippian Communities and Households*, ed. J. D. Rogers and B. D. Smith. Tuscaloosa, AL: University of Alabama Press, pp. 32–57.

Meltzer, D. J. (1993). Pleistocene peopling of the Americas. *Evolutionary Anthropology*, **1**, 157–69.

Meltzer, D. J., and J. I. Mead. (1983). The timing of late Pleistocene mammalian extinctions in North America. *Quaternary Research*, **19**, 130–5.

(1985). Dating late Pleistocene extinctions: theoretical issues, analytical bias, and substantive results. In *Environments and Extinctions: Man in Late Glacial North America*, ed. J. I. Mead and D. J. Meltzer. Orono, ME: Center for the Study of Early Man, University of Maine of Orono, pp. 145–73.

Meltzer, D. J., and Smith, B. D. (1986). Paleoindian and Early Archaic subsistence strategies in eastern North America. In *Foraging, Collecting, and Harvesting: Archaic Period Subsistence and Settlement in the Eastern Woodlands, Occasional Paper No.* 6, ed. S. Neusius, Carbondale, IL: Center for Archaeological Investigations, Southern Illinois University, pp. 1–30.

Mills, H. H., and Delcourt, P. A. (1991). Chapter 20: Quaternary geology of the Appalachian Highlands and Interior Low Plateaus. In *Quaternary Nonglacial Geology – Conterminous United States, DNAG Vol. K-2*, ed. R. B. Morrison. Boulder, CO: Geological Society of America, pp. 611–28.

Milner, G. R. (1986). Mississippian period population density in a segment of the central Mississippi River valley. *American Antiquity*, **51**, 227–38.

(1998). *The Cahokia Chiefdom: The Archaeology of a Mississippian Society*. Washington, D.C.: Smithsonian Institution Press.

Minnis, P. E., and Elisens, W. J. (eds.). (2000). *Biodiversity and Native America*. Norman, OK: University of Oklahoma Press.

Morrow, J. E., and Morrow, T. A. (1999). Geographic variation in fluted projectile points: a hemispheric perspective. *American Antiquity*, **64**, 215–31.

Morse, D. F. (1971). Recent indications of Dalton settlement pattern in Northeast Arkansas. *Southeastern Archaeological Conference Bulletin*, **13**, 5–10.

(ed.). (1997). *Sloan, a Paleoindian Dalton Cemetery in Arkansas.* Washington, D.C.: Smithsonian Institution Press.

Morse, D. F., and Graham, R. (1991). Searching for *Paleolama. Field Notes, Newsletter of the Arkansas Archaeological Society*, **239**, 10–12.

Morse, D. F., and Morse, P. A. (1983). *Archaeology of the Central Mississippi Valley.* New York, NY: Academic Press.

Morse, D. F., Anderson, D. G., and Goodyear, A. C. (1996). The Pleistocene–Holocene transition in the eastern United States. In *Humans at the End of the Ice Age: The Archaeology of the Pleistocene–Holocene Transition*, ed. L. G. Straus, B. V. Eriksen, J. M. Erlandson, and D. R. Yesner. New York, NY: Plenum Press, pp. 319–38.

Mosimann, J. E., and Martin, P. S. (1975). Simulating overkill by Paleoindians. *American Scientist*, **63**, 304–13.

Moy, C. M., Seltzer, G. O., Rodbell, D. T., and Anderson, D. M. (2002). Variability of El Niño/Southern Oscillation activity at millennial time scales during the Holocene epoch. *Nature*, **420**, 162–5 (Data archived at http://www.ngdc.noaa.gov/paleo/pubs/moy2002).

Muller, J. (1982). The Kincaid region. Unpublished paper presented at the Annual Meeting, Southeastern Archaeological Conference, Memphis, Tennessee, October 29, 1982.

Munson, P. J. (1984). Weedy plant communities on mud-flats and other disturbed habitats in the central Illinois River valley. In *Experiments and Observations on Aboriginal Wild Plant Food Utilization in Eastern North America*, ed. P. J. Munson. Indianapolis, IN: Indiana Historical Society, pp. 379–85.

(1986). Hickory silviculture: a subsistence revolution in the prehistory of eastern North America. Unpublished paper presented at the Conference on Emergent Horticultural Economies of the Eastern Woodlands, Carbondale, Illinois, 1986.

Nabhan, G. P. (1989). *Enduring Seeds: Native American Agriculture and Wild Plant Conservation.* San Francisco, CA: North Point Press.

(2000a). Native American management and conservation of biodiversity in the Sonoran Desert bioregion. In *Biodiversity and Native America*, ed. P. E. Minnis and W. J. Elisens. Norman, OK: University of Oklahoma Press, pp. 29–43.

(2000b). Interspecific relationships affecting endangered species recognized by O'odham and Comcáac cultures. *Ecological Applications*, **10**, 1288–95.

Newsom, L. A., Webb, S. D., and Dunbar, J. S. (1993). History and geographic distribution of *Cucurbita pepo* gourds in Florida. *Journal of Ethnobiology*, **13**, 75–97.

O'Neill, R. V. (2001). Is it time to bury the ecosystem concept? (with full military honors, of course!). *Ecology*, **82**, 3275–84.

O'Neill, R. V., DeAngelis, D. L., Waide, J. B., and Allen, T. F. H. (1986). *A Hierarchical Concept of Ecosystems*, Monographs in Population Biology 23. Princeton, NJ: Princeton University Press.

Overpeck, J. T., Webb, T., III, and Prentice, I. C. (1985). Quantitative interpretation of fossil pollen spectra: dissimilarity coefficients and the method of modern analogs. *Quaternary Research*, **23**, 87–108.

Owen-Smith, N. (1987). Pleistocene extinctions: the pivotal role of megaherbivores. *Paleobiology*, **12**, 351–62.

(1989). Megafaunal extinctions: the conservation message from 11,000 years B.P. *Conservation Biology*, **3**, 405–12.

Paine, R. T. (1966). Food web complexity and species diversity. *American Naturalist*, **100**, 65–75.

Parson, E. A., and Clark, W. C. (1995). Sustainable development as social learning: theoretical perspectives and practical challenges for the design of a research program. In *Barriers and Bridges to the Renewal of Ecosystems and Institutions*, ed. L. H. Gunderson, C. S. Holling, and S. S. Light. New York, NY: Columbia University Press, pp. 428–60.

Peacock, E. (1998). Historical and applied perspectives on prehistoric land use in eastern North America. *Environment and History*, **4**, 1–29.

Peacock, S. L., and Turner, N. J. (2000). "Just like a garden": traditional resource management and biodiversity conservation on the Interior Plateau of British Columbia. In *Biodiversity and Native America*, ed. P. E. Minnis and W. J. Elisens. Norman, OK: University of Oklahoma Press, pp. 133–179.

Pearsall, D. M. (1994). Issues in the analysis and interpretation of archaeological maize in South America. In *Corn and Culture in the Prehistoric New World*, ed. S. Johannessen and C. A. Hastorf. Boulder, CO: Westview Press, pp. 245–72.

Peebles, C. S. (1978). Determinants of settlement size and location in the Moundville Phase. In *Mississippian Settlement Patterns*, ed. B. L. Smith. New York, NY: Academic Press, pp. 369–416.

Perry, D. A. (1995). Self-organizing systems across scales. *Trends in Ecology and Evolution (TREE)*, **10**, 241–4.

Peterson, J. B., and Asch Sidell, N. (1996). Mid-Holocene evidence of *Cucurbita* sp. from central Maine. *American Antiquity*, **61**, 685–98.

Petruso, K. M., and Wickens, J. R. (1984). The acorn in aboriginal subsistence in eastern North America: a report on miscellaneous experiments. In *Experiments and Observations on Aboriginal Wild Plant Food Utilization in Eastern North America*, ed. P. J. Munson. Indianapolis, IN: Indiana Historical Society, pp. 360–78.

Pfeiffer, J. E. (1974). *Indian City on the Mississippi*. New York, NY: Time–Life Nature/Science Annual.

Pickett, S. T. A., Parker, V. T., and Fiedler, P. L. (1992). The new paradigm in ecology: implications for conservation biology above the species level. In *Conservation Biology: The Theory and Practice of Nature Conservation, Preservation, and Management*, ed. P. L. Fiedler and S. K. Jain. New York, NY: Chapman & Hall, pp. 66–88.

Pierotti, R., and Wildcat, D. (2000). Traditional ecological knowledge: the third alternative (commentary). *Ecological Applications*, **10**, 1333–40.

Platt, S. G., and Brantley, C. G. (1997). Canebrakes: an ecological and historical perspective. *Castanea*, **62**, 8–21.

Price, J. E. (1978). The settlement pattern of the Powers phase. In *Mississippian Settlement Patterns*, ed. B. D. Smith. New York, NY: Academic Press, pp. 201–31.

(1982). Mississippian subsistence in the Ozark Border area of southeast Missouri and the Vacant Quarter. Unpublished paper presented at the Annual Meeting, Southeastern Archaeological Conference, Memphis, Tennessee, October 29, 1982.

Ramenofsky, A. F. (1987). *Vectors of Death: The Archaeology of European Contact.* Albuquerque, NM: University of New Mexico Press.

Raup, H. M. (1937). Recent changes of climate and vegetation in southern New England and adjacent New York. *Journal of the Arnold Arboretum*, **18**, 79–117.

Rea, A. M. (1997). *At the Desert's Green Edge, An Ethnobotany of the Gila River Pima.* Tucson, AZ: University of Arizona Press.

Redman, C. L. (1999). *Human Impact on Ancient Environments.* Tucson, AZ: University of Arizona Press.

Richardson, D. M., Allsopp, N., D'Antonio, C. M., Milton, S. J., and Rejmánek, M. (2000). Plant invasions – the role of mutualisms. *Biological Review*, **75**, 65–93.

Rindos, D. (1984). *The Origins of Agriculture: An Evolutionary Perspective.* Orlando, FL: Academic Press.

(1989). Darwinism and its role in the explanation of domestication. In *Foraging and Farming: The Evolution of Plant Exploitation*, ed. D. R. Harris and G. C. Hillman. London, UK: Unwin Hyman Ltd, pp. 27–41.

Rindos, D., and Johannessen, S. (1991). Human–plant interactions and cultural change in the American Bottom. In *Cahokia and Its Hinterlands: Middle Mississippian Cultures of the Midwest*, ed. T. E. Emerson and R. B. Lewis. Urbana, IL: University of Illinois Press, pp. 35–45.

Risser, P. G., Karr, J. R., and Forman, R. T. T. (1984). *Landscape Ecology: Directions and Approaches.* Champaign, IL: Illinois Natural History Survey Special Publication No. **2**, 1–17.

Ruddiman, W. F. (1987). Northern oceans. In *North America and Adjacent Oceans during the Last Deglaciation, DNAG Volume K-3*, ed. W. F. Ruddiman, and H. E. Wright, Jr. Boulder, CO: The Geological Society of America, pp. 137–54.

Ruddiman, W. F., and McIntyre, A. (1981). The North Atlantic Ocean during the last deglaciation. *Palaeogeography, Palaeoclimatology, Palaeoecology*, **35**, 145–214.

Runkle, J. R. (1985). Disturbance regimes in temperate forests. In *The Ecology of Natural Disturbance and Patch Dynamics*, ed. S. T. A. Pickett and P. S. White. Orlando, FL: Academic Press, pp. 17–33.

Russell, E. W. B. (1983). Indian-set fires in the forests of the northeastern United States. *Ecology*, **64**, 78–88.

(1997). *People and the Land Through Time: Linking Ecology and History.* New Haven, CT: Yale University Press.

Sassaman, K. E. (1996). Early Archaic settlement in the South Carolina Coastal Plain. In *The Paleoindian and Early Archaic Southeast*, ed. D. G. Anderson and K. E. Sassaman. Tuscaloosa, AL: University of Alabama Press, pp. 58–83.

(1999). A southeastern perspective on soapstone vessel technology in the Northeast. In *The Archaeological Northeast*, ed. M. A. Levine, K. E. Sassaman, and M. S. Nassaney. Westport, CT: Bergin and Garvey, pp. 75–95.

Sassaman, K. E., and Nassaney, M. S. (eds.). (1995). Epilogue. In *Native American Interactions: Multiscalar Analyses and Interpretations in the Eastern Woodlands*, ed. M. S. Nassaney and K. E. Sassaman. Knoxville, TN: University of Tennessee Press, pp. 341–53.

Schiffer, M. B. (1975a). An alternative to Morse's Dalton settlement pattern hypothesis. *Plains Anthropologist*, **20**, 253–66.

(1975b). Some further comments on the Dalton settlement pattern hypothesis. In *The Cache River Archaeological Project: An Experiment in Contract Archaeology*, ed. M. B. Schiffer and J. House. Fayetteville, AR: Arkansas Archeological Survey, pp. 103–12.

Schoeninger, M. J., and Schurr, M. R. (1994). Interpreting carbon stable isotope ratios. In *Corn and Culture in the Prehistoric New World*, ed. S. Johannessen and C. A. Hastorf. Boulder, CO: Westview Press, pp. 55–66.

Schroeder, S. (1999). Maize productivity in the Eastern Woodlands and Great Plains of North America. *American Antiquity*, **64**, 499–516.

Schroedl, G. F., Davis, R. P. S., Jr., and Boyd, C. C., Jr. (1985). *Archaeological contexts and assemblages at Martin Farm*, Report of Investigations No. 39. Knoxville, TN: Department of Anthropology, University of Tennessee.

Schurr, T. G., and Wallace, D. C. (1999). mtDNA variation in Native Americans and Siberians and its implications for the peopling of the New World. In *Who Were the First Americans? Proceedings of the 58th Annual Biology Colloquium, Oregon State University*, ed. R. Bonnischen. Corvallis, OR: Center for the Study of the First Americans, Oregon State University, pp. 41–77.

Senft., R. L., Coughenour, M. B., Bailey, D. W., *et al.* (1987). Large herbivore foraging and ecological hierarchies. *Bioscience*, **37**, 789–99.

Sharp, W. E. (1997). *Management Summary of a Phase I Cultural Resource Survey of the Berea North Timber Sale and Associated Tracts, Jackson County, Kentucky, Berea Ranger District, DBNF, Investigations 96–13–MS3*. Winchester, KY: USDA Forest Service, Daniel Boone National Forest.

Simberloff, D., and Von Holle, B. (1999). Positive interactions of nonindigenous species: invasional meltdown? *Biological Invasions*, **1**, 21–32.

Smalley, G. W. (1986). *Classification and Evaluation of Forest Sites on the Northern Cumberland Plateau*. New Orleans, LA: USDA Forest Service General Technical Report SO-60, Southern Forest Experiment Station.

Smith, B. D. (ed.). (1978). *Mississippian Settlement Patterns*. New York, NY: Academic Press.

(1986). The archaeology of the southeastern United States: from Dalton to de Soto, 10,500–500 B.P. *Advances in World Archaeology*, **5**, 1–92.

(1987). The independent domestication of indigenous seed bearing plants in eastern North America. In *Emergent Horticultural Economies of the Eastern*

Woodlands, Occasional Paper No. 7, ed. W. Keegan. Carbondale, IL: Center for Archaeological Investigations, Southern Illinois University, pp. 3–48.

(1989). Origins of agriculture in eastern North America. *Science*, **246**, 1566–71.

(1992). *Rivers of Change: Essays on Early Agriculture in Eastern North America.* Washington, D.C.: Smithsonian Institution Press.

Stafford, T. W., Jr., Semken, H. A. Jr., Graham, R. W., *et al.* (1999). First accelerator mass spectrometry ^{14}C dates documenting contemporaneity of nonanalog species in late Pleistocene mammal communities. *Geology*, **27**, 903–6.

Stanford, D., and Bradley, B. (2000). The Solutrean solution: did some ancient Americans come from Europe? *Scientific American Discovering Archaeology*, **2**, 54–5.

Stoltman, J. B., and Baerreis, D. A. (1983). The evolution of human ecosystems in the eastern United States. In *Late-Quaternary Environments of the United States. Volume 2: The Holocene*, ed. H. E. Wright, Jr.. Minneapolis, MN: University of Minnesota Press, pp. 252–68.

Straus, L. G. (2000). Solutrean settlement of North America? A review of reality. *American Antiquity*, **65**, 219–26.

Struever, S. (1968). Flotation techniques for the recovery of small-scale archaeological remains. *American Antiquity*, **33**, 353–62.

Stuiver, M., and Grootes, P. M. (2000). GISP2 oxygen isotope records. *Quaternary Research*, **53**, 277–84.

Stuiver, M., Grootes, P. M., and Braziunas, T. F. (1995). The GISP2 δ^{18}O climate record of the past 16,500 years and the role of the sun, ocean, and volcanoes. *Quaternary Research*, **44**, 341–54.

Stuiver, M., Reimer, P. J., Bard, E., *et al.* (1998). INTCAL 98 Radiocarbon Age Calibration, 24,000–0 cal BP. *Radiocarbon*, **40**, 1041–83 (http://depts.washington.edu/qil/).

Surovell, T. A. (2000). Early Paleoindian women, children, mobility, and fertility. *American Antiquity*, **65**, 493–508.

Talalay, L., Keller, D. R., and Munson, P. J. (1984) Hickory nuts, walnuts, butternuts, and hazelnuts: observations and experiments relevant to their aboriginal exploitation in eastern North America. In *Experiments and Observations on Aboriginal Wild Plant Food Utilization in Eastern North America*, ed. P. J. Munson. Indianapolis, IN: Indiana Historical Society, pp. 338–59.

Tankersley, K. B. (1998). Variation in the early Paleoindian economies of late Pleistocene eastern North America. *American Antiquity*, **63**, 7–20.

Thompson, R. S., Whitlock, C., Bartlein, P. J., Harrison, S. P., and Spaulding, W. G. (1993). Climatic changes in the western United States since 18,000 yr B.P. In *Global Climates since the Last Glacial Maximum*, ed. H. E. Wright, Jr., J. E. Kutzbach, T. Webb III, W. F. Ruddiman, F. A. Street-Perrott, and P. J. Bartlein. Minneapolis, MN: University of Minnesota Press, pp. 468–513.

Tilman, D. (1982). *Resource Competition and Community Structure.* Princeton, NJ: Princeton University Press.

Torroni, A., Bandelt, H.-J., D'Urbano, L., *et al.* (1998). mtDNA analysis reveals a major Late Paleolithic population expansion from Southwestern to Northeastern Europe. *American Journal of Human Genetics*, **62**, 1137–52.

Turner, M. D., Zeller, E. J., Dreschhoff, G. A., and Turner, J. C. (1999). Impact of ice-related plant nutrients on glacial margin environments. In *Ice Age People of North America: Environments, Origins, and Adaptations*, ed. R. Bonnischen, and K. L. Turnmire. Corvallis, OR: Oregon State University Press and the Center for the Study of the First Americans, pp. 42–77.

Turner, M. G. (1989). Landscape ecology: the effect of pattern on process. *Annual Review of Ecology and Systematics*, **20**, 171–97.

Turner, M. G., and Gardner, R. H. (eds.). (1991). *Quantitative Methods in Landscape Ecology, Ecological Studies 82.* New York, NY: Springer-Verlag.

Turner, M. G., Gardner, R. H., and O'Neill, R. V. (2001). *Pattern and Process: Landscape Ecology in Theory and Practice.* New York, NY: Springer-Verlag.

Turner, M. G., Romme, W. H., Gardner, R. H., O'Neill, R. V., and Kratz, T. K. (1993). A revised concept of landscape equilibrium: disturbance and stability on scaled landscapes. *Landscape Ecology*, **8**, 213–27.

Turner, N. J., Ignace, M. B., and Ignace, R. (2000). Traditional ecological knowledge and wisdom of aboriginal peoples in British Columbia. *Ecological Applications*, **10**, 1275–87.

Urban, D. L., O'Neill, R. V., and Shugart, H. H. (1987). Landscape ecology: a hierarchical perspective can help scientists understand spatial patterns. *Bioscience*, **37**, 119–27.

Von Holle, B., Delcourt, H. R., and Simberloff, D. (2003). The importance of biological inertia in plant community resistance to invasion. *Journal of Vegetation Science*, **14**, 425–32.

Wallace, A. (1966). *Religion: An Anthropological View.* New York, NY: Random House.

Walthall, J. A. (1998). Rockshelters and hunter–gatherer adaptation to the Pleistocene/Holocene transition. *American Antiquity*, **63**, 223–38.

Ward, P. D. (1997). *The Call of Distant Mammoths: Why the Ice Age Mammals Disappeared.* New York, NY: Copernicus, Springer-Verlag.

Warner, B. G., Mathewes, R. W., and Clague, J. J. (1982). Ice-free conditions on the Queen Charlotte Islands, British Columbia, at the height of late Wisconsin glaciation. *Science*, **218**, 678–84.

Watson, P. J. (1989). Early plant cultivation in the eastern woodlands of North America. In *Foraging and Farming: The Evolution of Plant Exploitation*, ed. D. R. Harris and G. C. Hillman. London, UK: Unwin Hyman Ltd, pp. 555–71.

Watts, W. A. (1973). Rates of change and stability in vegetation in the perspective of long periods of time, In *Quaternary Plant Ecology*, ed. H. J. B. Birks and R. G. West. New York, NY: Wiley & Sons, pp. 195–206.

(1980a). Late-Quaternary vegetation history at White Pond on the Inner Coastal Plain of South Carolina. *Quaternary Research*, **13**, 187–99.

(1980b). The late Quaternary vegetation history of southeastern United States. *Annual Review of Ecology and Systematics*, **11**, 387–409.

(1988). Europe. In *Vegetation History*, ed. B. Huntley and T. Webb, III. Dordrecht, Netherlands: Kluwer, pp. 155–92.

Webb, T., III, Bartlein, P. J., Harrison, S. P., and Anderson, K. H. (1993). Vegetation, lake levels, and climate in eastern North America for the past 18,000 years. In *Global Climates since the Last Glacial Maximum*, ed. H. E. Wright, Jr., J. E. Kutzbach, T. Webb III, W. F. Ruddiman, F. A. Street-Perrott, and P. J. Bartlein. Minneapolis, MN: University of Minnesota Press, pp. 415–67.

Westley, F. (1995). Governing design: the management of social systems and ecosystems management. In *Barriers and Bridges to the Renewal of Ecosystems and Institutions*, ed. L. H. Gunderson, C. S. Holling, and S. S. Light. New York, NY: Columbia University Press, pp. 391–427.

Whitney, G. G. (1994). *From Coastal Wilderness to Fruited Plain: A History of Environmental Change in Temperate North America from 1500 to the Present.* Cambridge, UK: Cambridge University Press.

Whitney-Smith, E. (1998). Pleistocene extinctions: the death of an ecosystem (online publication at web site http://www.well.com/user/elin/extinct.htm/).

Whittaker, R. H. (1967). Gradient analysis of vegetation. *Biological Reviews*, **42**, 207–64.

(1972). Evolution and measurement of species diversity. *Taxon*, **21**, 213–51.

(1975). *Communities and Ecosystems.* New York, NY: MacMillan.

Whittington, S. L., and Dyke, B. (1984). Simulating Overkill: experiments with the Mosimann and Martin model. In *Quaternary Extinctions: A Prehistoric Revolution*, ed. P. S. Martin and R. G. Klein. Tucson: University of Arizona Press, pp. 451–65.

Wilkes, G. (1989). Maize: domestication, racial evolution, and spread. In *Foraging and Farming: The Evolution of Plant Exploitation*, ed. D. R. Harris and G. C. Hillman. London, UK: Unwin Hyman Ltd, pp. 440–55.

Williams, M. (1989). *Americans and Their Forests: A Historical Geography.* Cambridge, UK: Cambridge University Press.

Williams, J. W., Shuman, B. N., and Webb, T., III. 2001. Dissimilarity analyses of late-Quaternary vegetation and climate in eastern North America. *Ecology*, **82**, 3346–62.

Williams, S. (1977). Some ruminations on the current strategy of archaeology in the Southeast. Unpublished paper presented at the Annual Meeting, Southeastern Archaeological Conference, Lafayette, Louisiana, November, 1977.

(1982). The Vacant Quarter Hypothesis: a discussion symposium. Unpublished paper presented at the Annual Meeting, Southeastern Archaeological Conference, Memphis, Tennessee, October 29, 1982.

Wilson, M. C., and J. A. Burns (1999). Searching for the earliest Canadians: wide corridors, narrow doorways, small windows. In *Ice Age People of North America: Environments, Origins, and Adaptations*, ed. R. Bonnischen, and K. L. Turnmire. Corvallis, OR: Oregon State University Press and the Center for the Study of the First Americans, pp. 213–48.

Winterhalder, B. (1981). Optimal foraging strategies and hunter–gatherer research in anthropology: theory and models. In *Hunter–Gatherer Foraging Strategies*, ed. B. Winterhalder and E. A. Smith. Chicago, IL: University of Chicago Press, pp. 13–35.

Winterhalder, B., and Goland, C. (1997). An evolutionary ecology perspective on diet choice, risk, and plant domestication. In *People, Plants, and Landscapes: Studies in Paleoethnobotany*, ed. K. J. Gremillion. Tuscaloosa, AL: University of Alabama Press, pp. 123–60.

With, K. A., and Crist, T. O. (1995). Critical thresholds in species' responses to landscape structure. *Ecology*, **76**, 2446–59.

Woods, W. I., and Holley, G. R. (1991). Upland Mississippian settlement in the American Bottom region. In *Cahokia and the Hinterlands: Middle Mississippian Cultures of the Midwest*, ed. T. E. Emerson and R. B. Lewis. Urbana, IL: University of Illinois Press, pp. 46–60.

Wright, H. E., Jr. (1977). Quaternary vegetation history: some comparisons between Europe and America. *Annual Review of Earth and Planetary Sciences*, **5**, 123–58.

(1984). Sensitivity and response time of natural systems of climatic change in the late Quaternary. *Quaternary Science Reviews*, **3**, 91–131.

(1989). The amphi-Atlantic distribution of the Younger Dryas paleoclimatic oscillation. *Quaternary Science Reviews*, **8**, 295–306.

(1991). Environmental conditions for Paleoindian immigration. In *The First Americans: Search and Research*, ed. T. D. Dillehay, and D. J. Meltzer, pp. 113–35. Boca Raton, FL: CRC Press.

(1993). Environmental determinism in Near Eastern prehistory. *Current Anthropology*, **34**, 458–69.

Wright, H. E., Jr., Kutzbach, J. E., Webb, T., III, Ruddiman, W. F., Street-Perrott, F. A., and Bartlein, P. J. (1993). *Global Climates since the Last Glacial Maximum*. Minneapolis, MN: University of Minnesota Press.

Wyckoff, D. G., and Bartlett, R. (1995). Living on the edge: Late Pleistocene–Early Holocene cultural interaction along the Southeastern Woodlands–Plains border. In *Native American Interactions: Multiscalar Analyses and Interpretations in the Eastern Woodlands*, ed. M. S. Nassaney and K. E. Sassaman. Knoxville, TN: University of Tennessee Press, pp. 27–72.

Yarnell, R. A. (1976). Early plant husbandry in eastern North America. In *Cultural Change and Continuity*, ed. C. E. Cleland. New York, NY: Academic Press, pp. 265–73.

(1977). Native plant husbandry north of Mexico. In *Origins of Agriculture*, ed. C. A. Reed. The Hague, Netherlands: Monton, pp. 861–75.

Yesner, D. R. (1996). Environments and peoples at the Pleistocene–Holocene Boundary in the Americas. In *Humans at the End of the Last Ice Age: The Archaeology of the Pleistocene–Holocene Transition*, ed. L. G. Straus, B. V. Eriksen, J. M. Erlandson, and D. R. Yesner. New York, NY: Plenum Press, pp. 243–53.

Young, B. W., and Fowler, M. L. (2000). *Cahokia: The Great Native American Metropolis*. Urbana, IL: University of Illinois Press.

Zawacki, A. A., and Hausfater, G. (1969). *Early Vegetation of the Lower Illinois Valley*, Report of Investigations No. 17. Springfield, IL: Illinois State Museum.

Index

Page numbers are in normal type for text references and in *italic* type for references to figures.